Rebels for the Soil

Rebels for the Soil

The Rise of the Global Organic Food and Farming Movement

Matthew Reed

publishing for a sustainable future

London • Washington, DC

First published in 2010 by Earthscan

Earthscan Ltd, Dunstan House, 14a St Cross Street, London EC1N 8XA, UK
Earthscan LLC,1616 P Street, NW, Washington, DC 20036, USA
Earthscan publishes in association with the International Institute for Environment and Development

For more information on Earthscan publications, see www.earthscan.co.uk or write to earthinfo@earthscan.co.uk

ISBN: 978-1-84407-597-3 hardback

Typeset by Safehouse Creative
Cover design by Susanne Harris

A catalogue record for this book is available from the British Library

Library of Congress Cataloging-in-Publication Data

Library of Congress Cataloging-in-Publication Data

Reed, David, 1948-
  Escaping poverty's grasp : the environmental foundations of poverty alleviation / David Reed.
     p. cm.
  ISBN-13: 978-1-84407-371-9 (pbk.)
  ISBN-10: 1-84407-371-8 (pbk.)
  ISBN-13: 978-1-84407-372-6 (hardback)
  ISBN-10: 1-84407-372-6 (hardback)
  1. Poverty--Case studies. 2. Conservation of natural resources--Case studies. I. Title.
  HC79.P6R44 2006
  339.4'6--dc22

                         2006009672

At Earthscan we strive to minimize our environmental impacts and carbon footprint through reducing waste, recycling and offsetting our CO_2 emissions, including those created through publication of this book. For more details of our environmental policy, see www.earthscan.co.uk.

Printed and bound in the UK by TJ International,
an ISO 14001 accredited company. The paper used is
FSC certified and the inks are vegetable based.

Mixed Sources
Product group from well-managed
forests and other controlled sources
www.fsc.org Cert no. SGS-COC-2482
© 1996 Forest Stewardship Council
FSC

Contents

Preface

I have worked with food for most of my life, firstly in kitchens preparing and serving food and later as a sociologist who has specialised in researching rural areas. This book springs from that encounter of having been involved in feeding a mass of people and the need for a food system that takes into account the effects it has on people's health, the surrounding community and the environment. I have been fortunate to discuss this with chefs, farmers, growers, food processors, those who run supermarkets, scientists, activists, policy makers and consumers. All of them have shown an interest in the challenge that organic food and farming presents. Some of them react aggressively against it, decrying it as elitist, romantic and wasteful or as one farmer said pithily 'a bunch of wasters'. Others have been enthusiastic, looking to organics to produce healthier bodies, a more natural relationship between themselves and the planet, as well as tastier food. The response I never get is indifference; the debates sparked by the organic movement generate opinions and debate.

In this book I wanted to counter the common opinion that organic farming and food is inherently interesting or has only come to prominence because of geopolitics, a decline in rationality in our society or an increasing self-interestedness by consumers or the greed of supermarkets. Organic food and farming are part of our societal discussions because an assembly of people have put it there through their own actions. By understanding the way in which the organic movement operates as a movement operates, the history of its ideas and the networks through it has developed it becomes clearer that the prominence of the organic movement is a considerable achievement. It also begs the questions as to why is has not been more successful, and how it might be so. In making this argument I present evidence that will be uncomfortable for some in the movement, particularly around the history of the early years. Yet for those opposed to the organic movement it also poses the question of having to explain why the organic movement persists. Many groups and people criticise the status quo, few have a working replacement waiting in the wings to be introduced. The organic movement may not have all the answers, but it is posing important questions and has an alternative to hand. In a time looking toward a crisis in the continuity and supply of food, that must command our attention.

The research and thinking on which this book draws spans a decade so the thanks are wide ranging, but I would like to thank Peter Jowers, Harry Rothman, Dave Green and Alan Greer who were, or are still, at the University of West of England. At Exeter University Matt Lobley, Allan Butler and Mike Winter. For my brief but important time at the Open University special thanks are to due to Guy Cook for his encouragement and example. All of my colleagues at the Countryside and Community Research Institute have been, as they always are, an unfailingly supportive team, but particular thanks to Nigel Curry, James Kirwan and Paul Courtney. Thanks as well to Tim Hardwick and all the team at Earthscan for all their guidance, I'm still marvelling that I am an Earthscan author and many thanks to Carol Morris and the two anonymous referees for their constructive comments. I am blessed with a family who stand behind me without whom none of this would be possible and who have never shirked from encouraging me during the decade of research that stands behind this book. They inspired all that follows which is right all that errs belongs solely to me.

Tables and Boxes

Tables

Boxes

Acronyms and Abbreviations

BSE	Bovine Spongiform Encephalopathy
CAP	Common Agricultural Policy
CCOF	Californian Certified Organic Farmers
CSA	community supported agriculture
DDT	Dichloro-Diphenyl-Trichloroethane
EDAP	Energy Descent Action Plan
FAO	Food and Agriculture Organisation (UN)
FDA	Food and Drug Administration (US)
FIAN	Food First Information and Action
FSEs	field scale evaluations
FYM	farm yard manure
GAO	Organic Farming Group
GE	genetically engineered
GM	genetically modified
HDRA	Henry Doubleday Research Association
IARI	Imperial Agricultural Research Institute (India)
IFOAM	International Federation of Organic Agriculture Movements
IMF	International Monetary Fund
MST	Movimento dos Trabalhadores Rurais Sem Terra (Brazil)
NAFTA	North American Free Trade Agreement
NOP	National Organic Programme (US)
NPK	Nitrogen, Phosphorus and Potassium
OCA	Organic Consumers Association (US)
OTA	Organic Trade Association (US)
SCIMAC	Supply Chain Initiative on Modified Agricultural Crop
SMOs	Social Movement Organizations
UN	United Nations
UNDP	United Nations Development Programme
USDA	US Department of Agriculture
WTO	World Trade Organisation

1
Introduction

Quietly humanity has entered a new epoch; for the first time in history more people live in cities and towns than in the countryside. There was no ceremony to mark this transition; it made few column inches in newspapers and gained little mention on the TV networks. Yet the importance of this change appears to have become almost immediately apparent, as the international community has become consumed with concern about the future supply of food. Policy makers, think tank pundits and social activists point to the growing pressures of population, diminishing finite resources such as oil, water and phosphate as well as damage to the environment by pollution, that are all endangering the food supply. A spike in the price of staples such as rice, maize flour and wheat in 2007–2008 saw rioting by the urban poor in major cities in the south of the planet as their food supply suddenly became too expensive for their limited means. At the same time many people are living with the damage done to their bodies by too ready a supply of high calorie foods, which are often denuded by being over-processed. Those not haunted by the possibility of hunger are trying to moderate the excesses of a food system that delivers profits for a few multinational companies whilst damaging consumers' health and the communities they live in. Which technology to use to solve the environmental crisis of our times divides both public and scientific debate. Food and farming have become a social question in a new mode, with concerted discussions about the social and environmental consequences of how it is produced, distributed and consumed taking up a central place in many public discussions.

The global organic farming and food movement arose in an earlier ecological, environmental and economic crisis. During the late 1920s and early 1930s a diverse network of people, working separately in the British Empire and Germany, began to develop alternative answers to the growing ecological fears caused by soil erosion and the poverty of the diet eaten by most of the world's population. Even in the increasingly long shadow of Nazi ideas of blood and soil this network grew, inspired by the encounter of Western science with different societies and ecosystems. The post-war 'solving' of these problems with chemical based technologies and new ways of organizing agriculture saw the movement initially marginalized; it then returned with new arguments about agricultural pollution through the 1960s and 1970s. The initial networks of those people opposed to chemical fertilizers created a broader critique of Western models of agriculture that gained an ever-wider audience. The movement's arguments about the damage done by pollution to ecosystems and the humans dependent on them altered the terms of the debates about food.

The radicalism of the organic movement in the 1970s found a new form in the 1980s as it embraced ethical consumerism as a way of achieving change in the present rather than a yet unrealized future. This ethical consumerism saw the organic movement adopted by the emerging powers of the new food system, the multiple retailers or supermarkets. Organic food found its way into the lives and onto the dinner plates of more people than ever before, creating a burgeoning organic industry and government policy framework. As well as being marketed as an 'alternative' food product, the later part of this period saw the organic movement take to the field against a new technology – genetically engineered or modified (GM) plants. After losing the earlier debates about artificial fertilizers, pesticides and fungicides, the organic movement gained considerable territory in opposing GM technologies. This mobilization saw the movement transformed, simultaneously increasingly commercially successful and less comfortable with the strategy of growing through consumerism. The organic movement entered the economic crisis of 2008 uncertain of its strategy.

Many elements of these public debates have been constant, others are specific to times and places. Fear of ecological collapse, the damage inflicted by poor food and a crisis in the sustainability of farming are all familiar parts of the historical arguments in this global discussion, debates often sparked by the organic food and farming movement. The organic movement has been at the forefront of cultural discussions about the role of food in our societies and the consequences of its production for over 60 years. It has been consistent in questioning the reliance on oil-based chemicals in farming, the damage many technologies do the ecosystems underpinning agriculture, the over-processing of foods to remove their essential qualities, and the damage done to communities by the way in which they are then distributed. It has consistently argued that food should be grown in ways that enhance fertility through mimicking and enhancing natural cycles, that food should be a source of health and distributed in ways that nurture communities. This critique has seen the organic movement move from the margins to the centre of discussions about the future of food and farming.

At this moment, with the future of food and farming moving towards the top of the international agenda again, an analysis of the organic movement is necessary to chart how some of the alternative arguments to those that have come to dominate the debate have arisen. This book provides that analysis; it is not proffering the case for organic food and farming but rather an explanation of how that case has come to be made. It has four central arguments that weave into one another but can be separated for analytical purposes. First, the organic food and farming movement is a global social movement. This is a deceptively simple statement, but it means that there is no organic *farming* movement separate from an organic *food* movement, that it is planetary in its scope and always has been. As a social movement, we can understand it through theory in a way that helps us understand its behaviour and form. Second, the arguments of the organic movement have changed over time and space; they are not the same today as they were in the 1930s or as they were in the 1960s.

Many of the ideas proposed in the past would be an anathema to those proposing organic opinions today but they share a resemblance, not just in language but in both the critique levelled and in some cases the prescriptions made. Third, the organic movement is a cultural movement: it does not aim to win government or enshrine itself in law; rather it wants to change how we view and perceive our relationship to nature, and that in turn implies how we view one another. Fourth, it moves in waves or periods, in which it advances a particular set of arguments and organizes itself in specific form. It does this, borrowing a phrase from Jules Pretty, to change our collective relationship to 'agri-culture' (see Box 1.1).

Box 1.1 *Crop genetic diversity definitions*

As we advance into the early years of the 21st century, it seems to me that we have some critical choices. Humans have been farming for some 600 generations, and for most of that time the production and consumption of food has been intimately connected to cultural and social systems.   (Pretty 2002, pxii)

Jules Pretty in his book *Agri-culture* breaks the word into its two component parts, stressing the linkage between creating food as a physical system and the social and cultural activities around these processes. Pretty points out how the physical and cultural link around food production and consumption; that food production is not just a physical activity but also a cultural one that taps into a deep well of human culture. Equally, the cultural act of consumption is also a physical one; it ties us through the material we eat to those who produce it.

In considering the organic movement through the social sciences a number of new insights arise. The first and important step of seeing the organic movement as a social movement is to recognize that it has not been a cipher thrown around by the tides and flows of historical events but has had a hand in its own direction. Opponents of the organic movement will find incidents and arguments in this book that they may think refute the organic case. This book points to the persistence of the organic movement, its durability. Although often characterized as being anti-science and anti-reason it is far from that; often it is embedded in science. It represents an important constellation of arguments in our collective culture that need to be considered in any discussion of our shared agri-culture. People working together, organized across time and space, have achieved that position.

The organic movement has constructed a coalition of groups and networks that present not a theoretical plan but an active alternative. Since the 1930s they have put forward a set of ecologically redemptive technologies that have addressed the environmental question of the moment. In the 1930s it advanced composting to save the earth's fragile soils. From the 1960s it has been advocating food free from the toxic residues of pesticides and the pollution resulting from chemically intensive farming. In addressing climate change, it will call for both low-oil and carbon neutral at the very least. As well as a set of arguments about what is wrong with contemporary agriculture, the organic movement has at hand a set of technologies that it believes provide solutions.

This chapter sets out some of the basic ideas and arguments that are needed for the rest of the book, starting with a basic introduction to what distinguishes organic farming and food from the mainstream. It then discusses the four central arguments of this book, the evidence deployed and then provides a sketch of each chapter of the book by discussing the different 'waves' of the organic movement.

Organic farming 101

Although this is not a book about farming, but the arguments around it, it is necessary to establish what separates organic from other forms of farming. Organic farming is not a single set of farming practices or food stuffs, rather it is better thought of as a family of styles of farming that share some basic assumptions, often based on what they will not use or do. Most contemporary organic food stuffs carry labels or symbols that tell the consumer that the product has been grown without the use of artificial fertilizers, pesticides, fungicides, herbicides, prophylactic antibiotics, growth promoters, genetically modified plants and a range of food processing ingredients. Many of the chemicals and technologies are very common in most other foods, and alongside the claims about the absence of certain things will be a logo indicating that this lack has been certified by a third party. Most of the rest of this book looks at how those absences came about and rather less about what is done instead (Lampkin 1990).[1] This section provides a quick survey of how organic farming and food is different.

The first point of difference between organic and other farming systems is the rejection of artificial fertilizers, the type that is produced in large chemical factories and sold by the bag or even truckload. These fertilizers provide a ready source of the basic chemicals that help promote plant growth and have been widely used throughout the 20th century. By 'artificial', organic farmers and growers mean the inorganic chemical fertilizer NPK – Nitrogen (N), Phosphorus (P) and Potassium (the chemical symbol of the latter is 'K'). Alongside this, other secondary and trace elements can be added such as calcium, sulphur, magnesium, iron, copper, boron or chlorine. Already the language is contested over what is 'natural', with the chemical terminology of 'organic' and 'inorganic' adding to that debate.

Organic farmers make a range of arguments against these chemicals; that they do not promote growth in plants in the right way, forcing them beyond natural limits, so producing less healthy plants, so in turn less healthy animals and humans. This was the case made by the organic pioneers (see Chapter 3) before the Second World War. They also argued that such chemicals damaged the ecosystem of the soil, destroying its 'heart'. Later arguments were made about the damage to the wider environment as the chemicals can run off agricultural land and into watercourses damaging the health of those ecosystems and sometimes that of people, as in the case of methemoglobinemia or 'blue baby syndrome'. The arguments against fertilizers have moved from those specific to the crop and food, to include wider ones about pollution.

As agriculture is reliant on the fertility of the soil, adding nitrogen and other chemicals is fundamental to maintaining that fertility, but organic farmers use a wide range of alternatives to sacks of NPK. In part this defines the different sub-styles of organic farming, as each has a different approach. For most the basic material is animal manure or FYM (farmyard manure) that is collected and either spread onto the land or composted in some way before being taken back to the land. As we will see in Chapter 3, a husband and wife team by the name of Howard were early advocates of composting FYM before spreading it on the land. Biodynamic farmers, as well as capturing manure, also use other preparations to increase the fertility and, by the 1950s, a group of largely Swiss farmers had started to use volcanic dusts and rocks to add fertility to the soil.

The second way of adding fertility to the soil is through using plants to collect it. Several types of plants, for example some peas and beans, naturally collect nitrogen from the atmosphere and store it in their roots. Other plants capture different chemical elements, or send their roots down deeper to collect more minerals. By rotating crops or combining them, organic farmers seek to build the fertility of their soils. Some organic farmers have developed entire animal production systems on the careful management of permanent fields of grasses in combination with other plants or leys (Hassanein and Kloppenburg 1995). Others have created systems based on using these techniques to develop systems that are without animals, known as stockless.

Although seemingly arcane to the non-farmer, these differences are of profound importance, in that they govern the operation of the farm and so the production of food from it. For the pioneers, their ideal was the farm as a closed system, that produced all of its own fertility, and for some–for example biodynamic farmers – this remains a fundamental tenet of their farming. It also means that organic farms tend to be mixed farms, where a combination of crops and animals are produced, rather than the specialist farms that have become the norm in non-organic farming. These are tendencies, not inalterable rules, as some areas such as rugged hills, mountains or rangelands are only suitable for animal production, whilst some horticultural growers tend plots too small for stock animals. For those farmers using rotations and combinations to maintain and build the fertility of the farm, producing a diversity of produce rather than the same thing year after year makes it awkward to connect with the contemporary food system, which is based on uniformity. They have often found the answer to lie in customers who value that diversity and are prepared to support a farm by eating what is in season through a box scheme or community supported agriculture (CSA)(see Chapter 6).

The next absence, or set of absences, are pesticides or rather what organic growers and farmers call 'artificial' pesticides. These are also products of the chemical industry and they are used to kill insects, but we could also include fungicides and herbicides (aimed at weeds), which weaken or kill the intended crop or animal. The most infamous of these agents is DDT (dichloro-diphenyl-trichloroethane), first developed for use during the Second World War when the Allies wanted to kill mosquitoes and lice with it. After the war it was introduced for agricultural use and was used by public health authorities to clear malarial areas. DDT is a neurotoxin, which works along the insect's nervous system causing spasms and death. DDT was the first of a wave of new chemicals that targeted specific pests, fungi or weeds (Winston 1997). The idea of treating plants with chemicals to protect them and kill pests is hardly new. The organic movement argues that this technology is a step-change in that it kills so effectively it leads to ecosystems that are unbalanced. In addition, they point out that these chemicals persist in the wider environment, including in the human body (Beamont 1993).

Controlling pests and weeds can have a pronounced affect on the operations of a farm allowing greater concentrations of plants to be grown, improving the yield of some plants and making harvesting and storage easier. Organic farmers and growers were amongst the first to argue that they were also dangerous for humans, in that these chemicals occur nowhere in nature but are solely the products of laboratory processes. Their introduction into complex and planetary ecosystems would have unknown effects. They point to the role of these chemicals in human cancers, neurological disorders and birth defects. As many of the chemicals dissolve in fat they can be found in human body tissues and even breast milk. The use of these chemicals became part of the post-war changes to agriculture.

Organic farmers have a range of alternatives to such chemicals, largely based on those that can be derived from plant sources or were in use before the Second World War, although these are now complemented by newer technologies. The use of pesticides is minimized by growing combinations of plants – 'companion planting' – where one plant masks or discourages a pest or attracts predators of the pest. Some chem-ical treatments are used, for example, Bordeaux mixture is a combination of copper sulphate and hydrated lime used as an anti-fungal spray. Other hi-tech solutions are the use of cultured bacilli (see Chapter 7) or the introduction of insects that feed on certain pests. Instead of herbicides, organic growers use flame guns that use a short, intense burst of heat to destroy the cell structure of plants, or an old fashioned hoe (Blake 1994). Again this has implications for the farm as many of these technologies replace labour, and laborious work at that, as David Masumoto found on his family's Californian peach farm struggling against persistent weeds (Masumoto 1996). In California organic farmers found themselves battling to keep the short handled hoe in use as the state battled to outlaw it on worker welfare grounds (Guthman 2004). Labour is often one of the most expensive aspects of food production; increasing the amount of labour required pushes up the costs. Equally, organic growers and farmers may find that they produce less than their counterparts, also raising prices. As will be apparent as the book progresses, the importance and role of people working on the land is a constant theme of pro-organic arguments.

For the first 30 years of the organic movement the focus was on plants, as the health of animals was seen as being dependent on the quality of the food they received and their management. The inability to control disease meant that animals had to be kept in relatively low densities with the only real controls being careful husbandry and occasional drastic mass culls (Woods 2004). This situation was transformed after antibiotics were discovered and eventually approved for animal production, lifting the threat of epidemic and persistent disease. Animals could be confined and kept together in a manner that we have come to know as 'battery farming' after intensive chicken production, but also in feedlots for cattle or indoor farms for pigs. It was soon discovered that if antibiotics were given routinely, without any symptoms of disease, animals would grow faster. Alongside this new technology came the ability to synthesize growth hormones, which would mimic the animal's own and again boost the weight and growth of the animal. Animal production was transformed and intensive animal systems swept across the planet (Tansey and D'Silva 1999).

Organic farmers rejected these technologies on a number of grounds. Although not widely recognized now, many of the organic pioneers believed vaccinations and antibiotics would undermine health, arguing that the body should form its own resistance. This argument was not prominent in the organic movement in the 1960s, when antibiotics and growth promoters were the centre of concern, but it was still present. It wasn't long before those in the organic movement argued that human health was at risk, as animal diseases developed immunity to the antibiotics, an immunity that could threaten human health (Swann 1969). Also, it was suspected that a residue of the antibiotic could enter the human food chain, with much the same effect (Harrison 1964, 1970). As a following wave, arguments came that these practices were cruel and that they denied the animals the opportunity to behave naturally. Despite the potential benefits of cheap, high protein foods, this form of farming would produce meat that was denatured, lacking the qualities associated with an animal that had led a fuller life.

Different parts of the world regulate these technologies differently, for example, growth promoters are illegal in the European Union but not the US (see Chapter 7). Antibiotics similarly face different regulations, both in terms of what can be used and how they can be administered. Organic farming makes this complexity simple; growth promoters are not allowed and antibiotics are only for sick animals. This means that producing animals for meat, milk or eggs cannot be conducted at the same intensity as non-organic. Organically farmed animals are usually allowed to live longer to reach slaughter weight, so animal production is slower and there are fewer animals; once again the cost to the consumer tends to be higher.

Later in the book a whole chapter discusses the controversy around genetically engineered or modified (GE or GM) plants, but the arguments again tend to be about the natural behaviour of plants, the way in which pollen might pollute the rest of the environment and the unknown potential effects of this technology on human health (Chapter 7). A further aspect of this dispute is that many people opposed the domination of the technology by a handful of multinational companies, a critique that had also been levelled against pesticide production (see Chapter 5). The actual consequences of these technologies on farming practice for organic farmers and growers have been minimal but again genetic modification is banned under organic rules.

As may be apparent from this discussion, organic agriculture is now controlled by a set of rules about what is not allowed but not about what should be done. Exactly what needs to be done on a farm has to be devised by the organic growers and farmers to suit their particular farm, considering the soil, the weather, the surrounding ecosystem and their goals. This places an emphasis on the skill and ingenuity of the grower to understand the ecosystem of their farm, the behaviour of their animals and the needs of their plants. Over much of the 20th century agricultural science looked to make farming less particular, more controlled and calculable. Organic farming, in rejecting many of the technologies that have come to typify agriculture, has chosen to focus on another way of growing food and farming the land. Some organic growers and farmers can be said to have rejected technology, but most use a combination of the old and the new. One technology that no organic farmer can escape is certification – the rules of organic farming are discussed at greater length in Chapter 6, but inspection and working to agreed standards has become a central axis of organic farming and food.

All of these decisions can have consequences for the productivity of the farm, a controversial discussion point that has run for as long as the organic movement has existed. As with most measurements, this controversy revolves around what is counted, not just how it is counted. Productivity can be calculated on the unit of land used, the number of people employed or the amount of energy used to produce the food or, more recently, the quantity of greenhouse gases released (Pretty, et al 2000; Pretty et al 2005). Alternatively, the amount of damage to the wider environment or the benefits to the rural economy can be considered (Lobley et al 2005). Certainly the opponents of organic farming and food point to the importance of the production of food per unit of land used, focusing on the totality of the supply and quantity of food. In part this debate is a technical discussion; it is also a power struggle over the ability to define the problem and how it should be considered. It is this debate that forms the central cultural and social 'work' undertaken by the organic movement; Chapter 2 discusses this in further depth.

Box 1.2 *Organic vs conventional*

Often the debate between organic and non-organic food has been characterized as one between 'organic' and 'conventional' food systems. As the agricultural economist David Harvey observed, in one sense of the term organic farms are conventional in that they follow a strict set of rules whilst 'conventional' farmers have far greater discretion as to what they can choose to do. Rather then replay these opposites, it is more useful to investigate how these difference have been formed over time and in response to one another.

The discussion about productivity also turns attention to the role of science and technology in the organic movement. For some critics the organic movement is based on a rejection of science and a turn to magical or mystical formulations, captured in the phrase that it is all just 'muck and magic'. As is discussed in Chapter 3, this locution has more specific meanings, but it raises important issues. Some organic agriculture does rely on what many people would consider to be magic, the biodynamic preparations seek to capture 'astral' energies and many around the organic movement have a belief in the systemic consequences of very small quantities as in homoeopathy. For some organic farmers there are concepts of natural balance, that patterns or forms exist in nature, and that these are beneficial, or that God's work is apparent in them.

At the same time ecological science has a central role in how nature and farms are understood. Many in the organic movement have put considerable effort into scientific research and continue to see value in its findings, placing organic farming and food as a rational response to exuberant adoption of new technologies. Again there is no one organic position, rather a spectrum of opinions and arguments that are critical of the dominant technologies.

The organic movement has turned to science to verify these contentions and claims; much of the movement's efforts in the 1950s and 1960s was taken up with this strategy (see Chapters 4 and 5). Indeed it continues to search for credibility and the validation of science, which in some areas has been forthcoming, for example, in the levels of wildlife to be found on organic farms (Winter and Morris 2002). Other areas remain stubbornly resistant to such proof, and remain mired in controversy, such as the health claims around organic food. At times and in some discussions the organic movement has offered its own take on science. For example, some in the biodynamic movement use techniques to discover the auras around different foods and plants, which, they argue, demonstrate different facets of life. More frequently they argue that science is too narrowly focused and reductionist to encompass the interactions that make organic food and farming distinctive (Peters 1979; Lockeretz 2002). Rather they argue that totalities need to be considered, that science needs to become 'holistic' to provide a meaningful assessment of the organic case.

Organic farming and food can, therefore, be defined negatively by what it rejects and refuses to use or it can be discussed in positive terms. Organic farmers argue that their form of farming is more sustainable, paying attention to local conditions, and is less reckless, or not reckless at all, in the use of increasingly scarce natural resources. They argue that their way of farming allows for a greater diversity of wildlife on the farm and builds the fertility of the soils on which the rest of society depends. Organic food, it is similarly argued, is superior in that it contains more minerals and vitamins, that it tastes better and, in turn, is better for human health as it has not been degraded by industrial agriculture or food processing (Brown 1998). Finally, because of the certification arrangements around organic food and farming, in a world of adulteration, contamination and substitution, organic food can be trusted to be what it says it is (Duram 2005).

These disputes, claims and counterclaims about technology, science, nature and evidence are not necessarily as they appear – they are also attempts to take command of what counts as 'truth'. They are disputes about how we can and should know the world around us, and who has the authority to make such pronouncements (Ross 1994; Geiryn 1999). The organic movement, in refusing to accept the authority of some forms of science or in refusing some technologies backed by major corporations, has entered into a fundamental dispute about contemporary society. Rather than being a movement that is about science and technology per se, as I will argue in the next chapter, it is helpful to view it as a movement that is about changing our shared culture. At different times the positive or negative aspects of organic farming and food are at the centre of the debate, at others it is science or culture that is being disputed. The opponents of organic farming know and are able to attack the movement across the breadth of these ideas, often making considerable in-roads with their critiques. Yet they have to recognize that not only is the movement persisting, it is growing and by many measures it is flourishing.

Key arguments

Quite what organic farming is has been a matter of debate. For some people it is a farming system, for others it is a branding opportunity and for yet others it is an ideal or the programme of a pressure group. If we view it as a social movement that is focused on achieving cultural change we can respond to many of these debates and provide a way of placing organic farming within an analytical framework. Later there will be more detail on what a social movement is and how it behaves but, for the present, we can say that is a collection of people who subscribe to a set of ideas, who are prepared to undertake purposeful social action to advance their aims and who create a range of organizations to sustain and further not only their ideas but also to support their sense of togetherness.

Box 1.3 *Key arguments*

1　The organic food and farming movement is a social movement.
2　Its discourse and the arguments it has advanced have changed and mutated over time.
3　The organic movement is a cultural social movement, seeking to create change through how we think and debate the issues.
4　The organic food and farming movement has had distinct periods or waves of activity.

Viewing organic farming and food in this way has a number of important implications, the first of which is that we have to take the agency of the movement seriously. By agency we mean the ability to effect change rather than just be buffeted by the flows of society and history. This power is not unlimited, but the movement has some, and can determine how it wants to act in pursuit of its goals. Understanding the form of social power and influence that the organic movement can deploy is central to much of the rest of the book. Giving the organic movement a degree of power also brings into question to what extent it is responsible for its own relative success or failure.

　　Second, as this talk of goals, ideas and aims implies, there is a body of ideas, or a discourse that is present that simultaneously unites those in the movement and divides it from its opponents. This discourse is a collective creation; although there are some people who have an important role in articulating it or working on it, they are not its sole authors. Rather, their skill is in bringing ideas together and fusing them into novel assemblies or forms. This is not to deny their importance or acumen but rather to place it within the context of a flow of ideas, images and concepts that are reshaped and re-formed as events unfold. There is no single organic discourse, no 'true' organics to be found, but rather the organic discourse of a particular time or a specific group. As later chapters discuss, some of the ideas of the past are very different to those of the present.

A movement is also a collection of people who seek a mutual set of goals or pursue collective ideas. In the early years of the movement this was a relatively small number of people, with the very active and socially visible forming tight networks. Yet behind these very public faces there were always thousands, ranging later to tens of thousands and eventually millions of others taking a role, however marginal, in the movement. Famous individuals may have founded organizations, written books and made speeches, but it was the work of those in the background to these developments who were and are as important in creating the movement.

By a cultural movement I mean one that has sought to change the way in which a whole topic has been discussed, debated and thought about. Many movements seek political change, the amendment of a law or the removal of a government. The broad goal of the organic movement has been to change the way in which agriculture is discussed and viewed across the planet. It has proposed a different relationship between society and nature, between humans and ecosystems. To effect that cultural change it has argued through books, pamphlets and other texts, as well as demonstrating through farms, that other ways of practising agriculture are possible. This cultural campaign has seen the movement seek not only to become involved in the debate but to be the dominant way in which these matters are discussed.

The organic movement is a planet wide movement and has always aimed to be. This is a socio-spatial argument, that relationships between people have to be organized across space, how that is achieved and what impacts that has being significant. These influences can be logistical, given the problems of communicating across vast distances, but can also be cultural in that it may or may not be possible to relocate insights from one place to another so producing a new and hybrid set of ideas or practices. There has been considerable debate, both academic and popular, about globalization. The argument in this book is that what we are experiencing at the moment is the latest round of a series of developments that have bound the planet together in novel ways. Organic farming and food stems from an earlier round of this process, when the colonial administrators and scientists of the European empires discovered the rest of the planet. In particular it was the encounter between British imperial scientists with the people of the Indian subcontinent that created the cultural innovations that underpinned the formation of the movement.

This book is not an encyclopaedia of who did what and when in the organic movement across the planet over the last century; such a tome would be both impossible to write and very dull. I realize that I have missed out huge sections of the world, without accounting for their role and importance in the global movement. In particular I am aware that I have not included a discussion of the Japanese organic movement, the experiments in China or those on Israeli Kibbutz. Rather what is being made here is an argument using a set of analytical tools, the intention being to cover the major events and important trends. Most of the focus will be on the English speaking world starting with the UK, with a subsidiary set of illustrations and examples from the US, Canada, Eire, New Zealand and Australia, with discussions of the actions of people in India, China, Sri Lanka, Sweden, Germany, Switzerland, Italy, South Africa, France, Uganda, Kenya and Argentina. The organic movement was global from the beginning, in both its ambitions to change the world and to be organized across it.

As will become apparent, I will argue that the UK has had, at certain times, a central role in the global organic movement. The focus on Great Britain may seem a little imperialist – and it is, in that the British Empire, and the empire of science that in part underpinned it, were important in bringing together the influences that helped form parts of the organic movement, and its legacy went a long way towards founding the post-war movement. Later developments that were peculiar to the UK have also had an important role to play in the development of the movement. So it is not that there is anything particularly 'British' about organic farming and food, but that the flows of ideas, people and events have come together there. In the post-war period, when the US started to take on a global ambition, discussions appeared about advancing North American interests and science similar to those that took place in the pre-war period in Britain. As the movement grew during the 1970s, the number of national movements multiplied, each bringing their own particular trajectories and histories to those of the earlier movement.

It is also important to note at this early stage that this is an account of a farming and food movement. Often there is a focus, particularly in some academic literature, on the organic *farming* movement. Logically any movement in agriculture has to have consequences for food, either its quantity or quality or perhaps both, but also the organic movement for many years has made arguments explicitly about food, not just farming. There is also perhaps another reason; talking about food and the consumers of food is more difficult than talking about farmers and growers alone. Those who eat organic are harder to identify, more diffuse and dispersed than those who farm and historically they have been the subject of far less research. Considering farming and food together is the first step in thinking about organic farming and food as a social movement.

Many distinguished scholars of organic farming and food have found it difficult to account for aspects of the movement they are so familiar with. In part this is because they have frequently come from a scientific background so have not specialized in how and why social change happens, and on occasions because it is hard to tell the wood from the trees. Often people within the organic movement will refer to it as a 'movement' without following through exactly what that means. This book picks up that challenge to demonstrate how this way of understanding organic farming and food allows us to understand it more fully.

Organization of the book

The global organic movement has gone through four periods, or 'waves' that are distinct from one another. The first was in the 1920s and 1930s when there was no mass movement but a network of people investigating the ideas and themes that became the underpinning of a wider movement. Writers and scientists as diverse as Rudolf Steiner and Albert Howard, separately but simultaneously, started investigating the basis of agricultural fertility. This period ended with the eruption of the Second World War in Europe, but its influence is clearly apparent in the second wave of the organic movement (see Chapters 3 and 4).

The second wave of the global organic movement saw the foundation of the first organizations and the initial instances of a collectivity of people who shared the same goals and aspirations recognizing other people as sharing those ideas. This period from the mid-1940s through to the late 1960s saw organizations being formed, research conducted on a select group of farms and much innovative experimentation and thought. During this time the organic movement was marginalized by the success of the green revolution, but it formulated a powerful critique that gained a purchase during the 1960s (see Chapters 5 and 6).

The third wave started in the early 1970s when the global movement turned towards ethical or green consumerism backed by state and private regulation as the way of achieving the movement's broader goals. This is the movement that most readers will be familiar with – a fairly extensive collection of farms and producers, dedicated organic food chains and organic goods in supermarkets as well as specialist shops. Also during this period the first mass public protests led by the organic movement appeared, a radicalization not previously seen (see Chapters 6 and 7). The organic movement placed itself at the forefront of a mass public mobilization around a particular farming technology – GM plants. This series of protests brought the organic movement into a new coalition with the environmental movement as well as radical peasant and farmers' groups.

The fourth wave, as the final chapter of this book argues, is forming now around the question of 'peak food'. The second wave of the organic movement formed in response to the green revolution, and paradoxically during its third wave, it benefited from positioning itself in contrast to the success of that revolution. Now, during what may be its fourth wave, the terrain is being transformed and the organic movement, is mutating to meet that change. This fourth wave may lead to profound changes in the organic movement, as the green consumerism of the third wave gives way to new forms of building and sustaining a mass movement (see Chapter 8). In many ways the advent of peak food is the time that the organic movement has foretold and anticipated, yet that does not mean that it will see the movement achieve its goals.

Identifying organics

Organic food is a magnet to the news media, much ink is spilt and browser columns filled by opinion on organic food and farming, in part because it touches on some of the grand controversies of our times: the environment and technology or the relationship between humans and nature. Also it touches on the faddishness of food, the desire for bodies that are different to the rather too fleshy or rather too thin ones that we generally have and the constant stream of advertising about food to which we are subjected. The sociologist Barry Glassner in his book *The Gospel of Food* observes that most of our discussions and controversies see the food industry win, as they find another way to make money out of it (Glassner 2007). Glassner admits that he lost his initial prejudices about organic food but cannot bring himself to condemn big food companies because they ultimately provide people with solutions to their 'meal problems'. Glassner starts his book by citing studies that show that in blind taste tests people say they prefer foods labelled as 'natural' or 'organic', 'even when the researchers actually gave them conventionally grown foods' (Glassner 2007, pvii). He ascribes this to what psychologists call 'expectancy confirmation', that people tailor their experiences to what they expect to happen. The question Glassner forgets to ask is how did that happen, how specifically did organic achieve that status in people's lives?

Glassner makes a common error, of confusing the terms 'natural' and 'organic'. Often people use these terms loosely, having been encouraged to do so by marketing that finds advantage in creating confusion, as Glassner argues elsewhere in his book. The rest of this book will make clear that 'organic' has a distinct social trajectory that is not necessarily about food company experts. Rather it is the arc of a group of farmers, consumers, journalists, scientists, artists, troublemakers, horticulturalists and diverse others who have worked on many different societies and cultures to create that expectation Glassner reports. Nature has a role in this story, not only as a category of consumer items but as an active force, as do the people who have organized and discussed the importance of 'organic' food and farming.

Note

1   For a practical guide start with Nic Lampkin's *Organic Farming*.

2
Social Movements

Introduction

We cannot turn back. There are those who are asking the devotees of civil rights, 'When will you be satisfied?' We can never be satisfied as long as the Negro is the victim of the unspeakable horrors of police brutality. We can never be satisfied, as long as our bodies, heavy with the fatigue of travel, cannot gain lodging in the motels of the highways and the hotels of the cities. We cannot be satisfied as long as the Negro's basic mobility is from a smaller ghetto to a larger one. We can never be satisfied as long as our children are stripped of their selfhood and robbed of their dignity by signs stating 'For Whites Only'. We cannot be satisfied as long as a Negro in Mississippi cannot vote and a Negro in New York believes he has nothing for which to vote. No, no, we are not satisfied, and we will not be satisfied until justice rolls down like waters and righteousness like a mighty stream. (Dr Martin Luther King 1963)[1]

Many of us can conjure up a mental image of Dr Martin Luther King making his 'I have a dream' speech, not just the power of the oration, not just the haunting irony that he did not make the promised land, the Shakespearian cadences, the preacher's call and response but also the tens of thousands of people looking on. People marched to the capital city, singing and carrying placards; their collective identity was obvious – not just black people but others who could clearly see the injustice they were subject to. What they all wanted was legal change, as well the end of racism and its institutions. This was the most analytically transparent of social movements, the cause was and is obvious, the objectives clear, and the institutions and tactics rather straightforward, their songs and cultural demands are still very close to the surface of much of Western culture. Unfortunately for those interested in social movements, not all movements are quite so easy to interpret, nor their victories and failures quite so well known.

This chapter is about how we can understand social movements, and the organic farming and food movement in particular. That we can even describe a movement about food and farming as being in some ways similar to one dedicated to the removal of racism signals both the power of the approach as well as the problems with it. In this chapter I set out the defining characteristics of a social movement, and how these need to be understood and modified to be applicable to the movement we are discussing. This takes the focus of our discussion away from social movements as always being concerned with politics and physical resources, towards arguments about the importance of culture and the fulfilment of the individual. In doing so a wide range of theory is discussed, enough hopefully to reveal the features of the movement but not so much as to leave the reader stuck in the swamps of too much theory.

Why social movements?

In everyday parlance we use the term 'movement' as verbal shorthand for social movement – there are animal rights movements, women's movements, peace movements, environmental and ethnic movements. Our civic, political and often personal lives are cut across by a multitude of movements, all of them sharing some aspects that are similar but also distinct. For many commentators these movements have become one of the defining characteristics of our age. Manuel Castells, as part of his opus *The Information Age*, argues that social movements are the way that people together challenge 'globalization and cosmopolitanism on behalf of cultural singularity and people's control over their lives and environment' (Castells 1997, p2). He argues that movements are not inherently good or bad, reactionary or progressive, but rather they are 'symptoms of who we are', and he draws our attention to the importance of the word 'us', as social movements are about how 'we' define each other and ourselves. The plurality, the wide array of social movements is not accidental but a sign of how we are trying to work out the world around us, seeking to control and contest parts of it. Castells, in his discussion of movements, highlights those as diverse as the militias in the US and the global feminist movement.

As Castells suggests, social movements are important as they capture something of the flow of a society, they are in movement and reflect as well as cause change. They allow us, as Nick Crossley argues, 'to gauge the workings of the broader political structures of our society' (Crossley 2002, p9). Social movements can be ways of understanding wider tensions within a society and how they are being played out. So in thinking about social movements we are also invited to think about the big themes of any period in history. Whilst this might suggest to us that social movements are important, holographic fragments of a broader story, that still leaves the problem of understanding the 'how' and 'why' of social movements.

Defining a social movement

Much ink has been spilt over how to define a social movement with several broad approaches to the question that are often characterized as the differences between those adopted in the US and in Europe. These continental labels are very broad and there are exceptions in each case but there is some validity in the observation. Before the 1970s social movements tended to be grouped with other questions of collective behaviour in the US; this shifted during the 1970s to questions about how people used the resources available to them and engaged with the political process. This approach, often called Resource Mobilization Theory, was focused on the 'how' of social movements, how they worked, how they were organized, who took part and how that could be understood. In contrast academic Marxism, considering particularly the privileged place of workers' movements, dominated the European model before the 1970s. During the 1970s this was contested and largely overtaken by approaches that emphasize the importance of culture. Broadly, European social movement theory has been concerned with the 'why' of social movements, why they form, why people take part and what they mean for broader society as well as daily life (Della Porta and Diani 1999).

Rather than plunge into a tour of these differences, and the academic history of them, in this book I focus on approaches from scholars who have done a great deal of work to try to bring the accounts of the 'how' and the 'why' closer together. As should become clear, the focus of this method allows a greater understanding of the organic food and farming movement, its successes and limitations. This approach focuses less on the use of protest and engagement with politics and more on the work done by social movements in experimenting with other ways of living. Much of the activity of a social movement happens 'below the waterline'; it is outside the public sphere, only breaking into plain view during moments of change or protest. Just because a movement is not on the march does not mean that it is inactive, it may be at its most creative during this period. At the same time some attention needs to be paid to the 'how' of social movements; organization plays an important role and the organic movement is no exception (Melucci 1985, 1989).

Donatella Della Porta and Mario Diani have done a great deal of work to reconcile the North American and European traditions, to which I will introduce some of the work of their colleague Alberto Melucci, as well as Nick Crossley and Thomas Rochon (Rochon 1998; Della Porta and Diani 1999, 2006). Della Porta and Diani have developed a four-part definition of a social movement that comprises 'informal interaction networks', 'shared beliefs and solidarity', 'collective action focusing on conflicts' and 'use of protest'. Each of these needs some description and qualification, but already the contours of their approach can be seen (Della Porta and Diani 2006).

Informal interaction networks

Social movements are very horizontal, without a great deal of hierarchy or indeed the internal structures that might characterize a large business or government department. Although it may contain organizations that are part of the movement, these have no particular primacy and ad hoc groups or networks may be more important. There are often few formal leaders, no membership cards for the wider movement and openness to dialogue and debate. The networks come in a variety of forms, with some being very tightly clustered, everyone in them knowing one another, but others are loose and diffuse. More often there are mixtures of networks; some groups of people are closely bound together whilst others are very loose, with a few people acting as links between the different groups. This flexible internal structure makes a social movement both very robust and highly adaptable but it can also make it slow to respond to opportunities; it can be very difficult for social movements to achieve particular goals. This means that at times specific organizations that are part of a wider movement carry out specialized and difficult tasks, leaving the rest of the movement amorphous and loose.

Shared beliefs and solidarity

Sharing some beliefs is the basis of a social movement, which in turn brings a sense of belonging and a willingness to act in support of one another. It is worth pausing to consider the implications of that solidarity as it means that movement participation is not about class, social position or status but a willingness to engage with and accept a set of beliefs. The result of this is that movements are often a social mix of people who would not normally be expected to find common cause. For some observers this looks like a fixation on single issues, questions of identity or lifestyle that do not add up to a political platform as understood in the political system. It also means that movements do not act on behalf of a social class, a frustration to many earlier Marxist commentators.

A sense of belonging is not a small effect of movement membership, but a profound one. Recognizing others as being members of the same movement, that they constitute part of the 'we', is often an intense personal experience that changes people's lives. People act in solidarity with one another recognizing that they share similar or identical goals that overcome other social boundaries such as class, geography or sometimes other political affiliations. This sense of belonging often means that individuals remain within the networks of the movement for much of their lives, providing the movement with a great deal of continuity over time (Polletta and Jasper 2001).

The other aspect of shared beliefs is not just to identify problems but also solutions. People may enter a movement because of identifying something that they do not like or agree with, but the locus of a movement is around the opening up of possibilities. Much of the activity of the movement is about exploring other answers and alternatives to the dominant answers in social life. In that way radical ideas such as equality between ethnic groups or preparations for a millennial religious conflict can be explored. This work is often done at the level of daily practice, living differently in anticipation of the wider social changes the movement wants to introduce, but also at the level of language and symbols. Social movements try to change how ideas are understood as part of their work of social change; protests may be targeted at symbols or the context in which they are understood may be altered (Tarrow 1996; Tilly 2004).

Collective action focused on conflict

Social movements are involved in conflicts. They are contesting with other social groups a social stake that both view as important; some movements are trying to change that social stake and others to protect it from change. The organic food and farming movement is trying to change the technology and social arrangements around agriculture, a fundamental activity in any society. This conflict can take a variety of forms but it always leads to claims and counterclaims, as each group feels that they have a lot to lose in this dispute. Each group may define itself as very different to the other but they are locked together because of the dispute, unable to be disentangled from one another. Recognizing and reacting to opponents is a very important part of the activities of a movement, introducing a dynamic that is not just about their internal debates or the movement of wider societal events. Conflict can change not just the stake in question but also movements and opponents alike.

Use of protest

Protest, which is usually defined as unconventional participation in politics, has long been seen as *the* defining characteristic of a social movement. This argument holds that it is through street protests, non-violent direction action, publicity grabbing stunts and mass lobbying that the real difference of a social movement is demonstrated. Yet, many of the scholars of movements have argued, much of the activity of a social movement is focused on cultural or social change and this would be missed by an approach centred on public protest. Also conventional politics has learnt to ape the tools of the social movement; mass protests or publicity stunts are not the sole preserve of movements but may also be the mark of the political lobbyists or corporations. It has also been noted that movements go through long periods of apparent inactivity and then suddenly become involved in public protests. So protest is part of social movement activity, but it is not the defining characteristic in quite the way often argued in the past.

Scholars who study the role of politics in everyday life have also pointed to the importance of micro-politics. The decisions that people make in their daily lives can have consequences that resonate more widely. During the campaigns to abolish the slave trade it was possible to buy sugar guaranteed not to have been produced on plantations that used slaves, and the boycott has become a tool that many institutions have faced since that time. These decisions are not only important when they are aggregated across a large group of people doing the same thing but they are also expressive of a person's identity. This is of direct relevance to movements like organics where one of the strategies is to encourage a particular form of consumption and follow through the consequences of that change. People are normally supposed to select products either because of advertising or the use value of a product, but to choose it because of its relation to broader social and political goals breaks with the logics of the consumer society. It is not a qeoture but a private one, which – when conducted in the context of a social movement – can become a form of protest (Michelleti and Stolle 2007).

The four-part definition by Della Porta and Diani covers most of the central features of a social movement, allowing us to see both the complexity of a movement and also what is shared between them even in their diversity. It also brings to the fore some of the questions that we need to consider when thinking about the organic food and farming movement. First, the organic movement has few expressly political goals, it is not looking for laws or particular policies rather it is about a much broader change. Second, this means that we have to consider the role of culture in change, how norms and values are reshaped in society and who does this, which means looking at the role of ideas and their place within movements. Third, from this we need to consider how a movement is organized across space and through time in order to achieve those goals. In much of the literature on social movements those that are explicitly concerned with cultural change are noted but rarely analysed in any depth because of their complexity and, in being less oriented around protest, they are less public.

Critical communities and discourse

Thomas Rochon, in his work on cultural change, argues that social movements are simultaneously active in two arenas, the social and the political (Table 2.1) (Rochon 1998). This characteristic of movements separates them from pressure or lobby groups, which focus on the political. Although in Table 2.1 it would appear Rochon favours public protest in both the political and social arenas, in the text he is clear that: 'The social arena is the world of changing values, identities, preoccupations, and daily behaviours' (Rochon 1998, p33). Of course these arenas cannot be neatly divorced in reality but it does aid analysis, particularly when considering the changes made by culturally inclined social movements. Political gains tend towards being most effective in the short term; they can be achieved and seen to be so, marking institutional and formal shifts in a society. Social and cultural change often runs before such changes, with legislatures and elected politicians confirming social changes that have been underway for some time – the results of the indefinite social campaign of a movement.

Table 2.1 *Two arenas of movement activity after Rochon*

Traits	Social arena	Political arena
Seeks to influence	Cultural values	Government policies
Target of activity	Individuals	Institutions
Optimal organization	Decentralized, participatory	Centralized, hierarchical
Typical activities	Teach-in, door-to-door canvas	Lobbying, referendum, electoral organizing
Time horizon	Indefinite	Bounded campaigns

Source: Rochon 1998, p33.

Rochon also poses the important question of where the ideas underpinning the goals of a social movement come from. The networks and structures of a movement are not the usual means through which innovative ideas are generated, even if they are excellent at diffusing them. This role of innovation, Rochon suggests, is conducted through 'critical communities', who together reconceptualize social questions, literally thinking about things in new ways. Rochon argues that these communities are 'networks of people who think intensively about a particular problem and who develop over time a shared understanding of how to view that problem' (Rochon 1998, p25). Those people in these communities are critical of the existing order of ideas and policies, but are not necessarily well connected to policy making and are not about making things more certain, rather they are interested in making them less so. Rochon argues that these communities become intertwined with social movements, which act to filter their ideas and diffuse them more widely.

This places one of the central roles of movements as changing the way that we think about social questions, it – as Melucci describes – a challenging of the 'codes' that we use to recognize and organize our society (Melucci 1996a). The concrete way in which these reconceptualizations and recodings are expressed is through language; social movements attempt to change the way that written and visual language discusses particular problems. As a shorthand way of discussing this package of metaphors, concepts and words, the term *discourse* is very useful:

> *The linguistic expression of a system of thought is called a discourse: a shared set of concepts, vocabulary, terms of reference, evaluations, associations, polarities, and standards of argument connected to some coherent perspective on the world.* (Rochon 1998, p16).

Although it is difficult to trace in the abstract changes in culture, because it is so diffuse and decentralized – as will be demonstrated in later chapters – it is easier to track the changes and evolution of a discourse.

Rather than follow Rochon in viewing critical communities as external to a social movement, I argue that most movements now contain them internally, often in an institutionalized form. Again, analytically, the separation between a movement and a critical community may be useful, but it is often hard to distinguish in events, as those central to a movement may also play a role in the community's recoding how an issue is conceptualized. It does remain a useful insight into how movements form, and the central role of cultural and intellectual change inherent in a social movement. Fighting in the street, or getting arrested dressed as a rubber chicken, may be the most visible manifestations of a movement, but its social and cultural work is conducted elsewhere and in a different form.

This broad cultural change is rooted in the experience of the individuals taking part; participation is rewarding and often transformative for those involved.

> *The joyful and encouraging discovery that other people are living the same experience is a strong support for individual commitment to collective action. Mutual affective recognition is not a post factum event but a central factor in the process of involvement.* (Melucci 1996, p301).

As the quote above suggests, there is a lot of emotion in taking part in a movement, and that affective element is part of creating and sustaining the networks that make up social movements (Jasper 1998). Those who become involved in a movement are not those on the margins or those who are alienated from society, but those who are well connected to it. In part this is because those who are integrated into society are part of the networks through which ideas circulate and have an idea of their own ability to make a change, what sociologists call 'personal agency'. Most people are recruited into movement networks by those they already know; friends and associates with whom they share at least some bonds of trust that they carry into the new movement. This means that the networks are often made of circles of friends and associates, who may share more than the movement in common.

Box 2.1 *Eve Balfour the protestor*

Eve Balfour pioneered women's role in farming, jazz, writing detective novels and aviation but came to the notice of people in the UK during the Tithe Tax protest of the late 1920s and early 1930s (see next chapter). The tax was paid to the Church, a hangover from an earlier period, and was at first bitterly resented by the small farmers of England. This resentment became refusal during the inter-war farming recession, with the Church using the law to enforce sales of farmer's property to recover the tax owed. Balfour was amongst the refusers and encouraged others to do so, gaining her a high profile amongst farmers. It brought her into contact with a wide range of political groups, but most consistently with Sir Stafford Cripps, a socialist aristocrat who became the Chancellor of the Exchequer in the post-war Labour government. Balfour brought the experience of the Tithe Tax into the organic movement.

Having some grounding in existing networks means that experience of other movements is often brought into a new movement, therefore movements do not have to learn from scratch. This can be an advantage but can also have some drawbacks (McAdam 1986). In Chapter 3 we will see how the Soil Association appeared out of the networks of the 1930s in which participants learnt a great deal about political activity. Together those taking part in the early organic movement constructed a collective identity, one that defined what being in the movement meant but also brought people into a common frame of reference. This shared identity is not a fixed thing but a field of possibilities within which people operate, allowing it to change over time, as tensions appear within it and the external environment alters. In Chapter 5 it becomes very clear that during the early 1970s the collective identity of the organic movement was profoundly challenged and this led to a process of transformation.

Phases of the movement

Movements are not active all the time, or rather they do not appear to be active. Melucci noted that movements go through periods of latency when they are not protesting but are engaged in cultural work, prefiguring the outcomes they would like to see in the future (Wall 1999). This is an important feature of contemporary movements, that they do not delay implementing the changes they would like to see until the conditions are right or some victory achieved but they attempt to realize them in the present. People seek to realize their needs in the present with their actions acting as an example of what they would like to achieve but also an experiment in what solutions are possible. At times this can appear to others as ridiculous, the subject of mockery to those who are opposed to their ideas, but over time becomes an established part of a society's wider life. This distinguishes social movements from revolutionary groups who are looking for a profound change and also from those who are looking solely for a change in government policy.

The idea that movements are still active when not engaged in public protest links to the idea that there are waves of mobilization in the career of a social movement, that there are moments of political opportunity or social change (Melucci and Avritzer 2000). The same movement can pass through distinct phases linked to particular events, changes in the collective identity or challenges to the movement. In feminism the protests around suffrage in the early 20th century were followed by protests around social and cultural equality in the later 20th century. Whilst the animal rights movement focused on anti-vivisection in the late 19th century, it shifted to the trade in feathers in the very early 20th century to return to anti-vivisectionism in the late 20th century. These waves of protest are the public face of these movements, telling us nothing of the internal debates and developments that led to those mobilizations (Tarrow 1996; McAdam et al 2001).

In a movement with a prolonged existence, it is unlikely to maintain its activity at a constant level but for it to ebb and flow, either because the circumstances in which it operates change or through developments internal to the movement. The global organic movement has gone through three waves or phases and is entering into a fourth now. This first period of activity is the prehistory of it as a movement, when it existed largely as a network of people scattered across the globe who were in pursuit of shared interests and topics with an increasing sense of being part of a collective effort – a critical community (see the next chapter). This changed in the second phase of the movement, which reached its peak in the late 1940s when for the first time the movement came together with a shared sense of being a group; the 'we' emerged and brought people together in shared institutions (Melucci 1996b). Although it had peaked by the late 1940s, it produced institutions and the impetus to create a constellation of farms that were experimenting with organic farming, looking to prove its superiority and to how it could be codified (see Chapters 4 and 5).

Table 2.2 *The first three waves of the global organic movement*

Phase	1st phase	2nd phase 1945–1969	3rd phase 1971–2005
Arguments	Soil science and rural communities	Based in soil science and agriculture	Environmentalism – pollution in the wider environment, contamination of food, conservation
Strategy	Discussion and writing	Scientific proof of superiority of organic farming and food	Practical experimentation to create viable system. Develop a market for organic produce
Tactics	Critical community	Research stations and demonstration farms/plots	Develop a practical farming system and low-level consumption. Develop institutions to support the market
Role of participants	Discussion	Financial support, gardening and dis-cussion	Consumerism, and discussion

As this second phase decayed in the late 1960s, a second wave rose along with the wider environment movement. This new wave of the movement emphasized environmentalism in various forms, and shifted the emphasis of the arguments within the movement towards the pragmatics of achieving organic production (see Chapters 5 and 6). In turn this meant that the strategy of the movement began to be placed in a code as to how to be organic in production and how that might be sold to those who were interested. Most members of the movement were still expected to garden to get their organic produce, but some consumption would be possible. This phase of the organic movement became very influential in the wider environmental movement, acting as a critical community beyond the boundaries of the organic movement (Pepper 1984). The discourse developed by Fritz Schumacher, Teddy Goldsmith and others clustered around the Soil Association in the early 1970s re-defined environmentalism (see Chapter 5). This was overtaken by a fourth and more aggressive phase that came from within the organic movement, that wanted not only to be conflictual in taking on its opponents but also placed ethical production and consumption as the key strategy of the movement. Participants would be asked to purchase organic produce and in doing so use the market as a way of making the movement prominent. In order to support that effort, the organizational underpin-ning of the movement would have be overhauled (see Chapters 6 and 7). As I will suggest in the concluding chapter, this is now being overtaken by a new wave of the movement adapted to the changing circumstances of the present (see Chapter 8).

This periodization is a way of thinking about the movement developing rather than a grid that can be laid across any particular example as a guide to what was happening at any one time. Yet, as later chapters will demonstrate, it does provide a way of analysing the development of the movement, and the dynamics within it. Some of these changes were given impetus by factors external to the movement, as it sought to take advantage of the situation, but others were often internal, as tensions and conflicts were resolved. Being an amorphous movement rather than a sleek organizational weapon, there is often a lag between an external event and a reaction, just as internal changes take time to filter through. This means that sometimes the patterns are geographically and temporally complex, some parts are catching up with the trend whilst others are innovating away from it. As is often the case, it is easier to plot these differences in retrospect than it is when they are taking place.

Discourse and diffusion

Sidney Tarrow argues that with the rise of the popular press came the rise of popular protest, that many social movements have been forged through the weak ties of print rather than the supposedly stronger bonds of class or geography (Tarrow 1996). In any movement that is planetary in its scale and ambition, how it is diffused or spread is a matter of central importance. The first and most obvious way in which movements are diffused is through the mobility of people who talk to and exchange experiences with others. Later chapters show that several key people in the early networks and years of the movement were very geographically mobile, as well as working in organizations that had global ambitions. Through a combination of economic privilege or professional status, they were able to travel between the continents, making friends, paying visits and giving talks as they went.

Equally, we can also see that people were able to correspond across large distances, there was an exchange of letters between England and the US, England and Germany, as well as many other personal correspondences. These became the loose links through which experience was exchanged (see Chapter 3) that did not allow for the coordination of daily activities but of organizational forms or innovations that might have purchase elsewhere. So Jerome Rodale was inspired to investigate organic agriculture after exchanging letters with Albert Howard, and Walther Darré's exchange of letters with Rolf Gardiner to attempt to found a German version of the Soil Association (see Chapter 4). Just as with the advent of cheaper telephone calls, and more recently information technology, one to one exchanges are still conducted. Whilst these contacts are important to those touched by them, they are not the basis of a mass movement.

The tools of mass communication, until very recently, have been around print media, not just pamphlets, newsletters and magazines, but also books and reports. Print media, although apparently flimsy, is very durable and easily transportable, capable of being sent relatively cheaply and reliably across the planet. Print has often brought together what Tarrow describes as 'invisible communities' (Tarrow 1996, p51), and for the organic food and farming movement in the first years the main focus of this nascent community was around books. At several key points in the development of the movement books have played a key role in changing the focus of not just campaigns but the whole movement. Unlike magazines or newsletters, these books were attempting to address an audience both inside and outside the movement. This is not to suggest that the organic food and farming movement is a literary association, or one that is dominated by those who can make it to print, but that books and print matter in the career of a movement.

Many movements use symbols to indicate either their presence or their goals; the scrawled red A in a circle for anarchism or the symbol of the peace movement are familiar as graffiti in many urban areas. Activists in movements may sport badges, bumper stickers or particular forms of clothing to signify their affiliations. These symbols act as social shorthand to those within the movement as well as those outside it, reinforcing the shared identity of the movement. Symbols are displayed at meetings and public protests; their prevalence, even ubiquity, is a guide to the movement's status and success. The diffusion and formulation of symbols in a movement is an important activity and one that has a central place in the organic food and farming movement. This is important in considering the rise and role of organic consumerism in the 1980s (see Chapter 6).

Roles in the movement

Some authors have written of the organic farming movement as if it were a separate entity to the wider organic food and farming movement, which foregrounds the importance of farming and farmers (Lockeretz 2007). Social movement theory allows us to see that an isolated focus on farmers and farms considers only the most active aspect of the movement. It concentrates only on those who are able to make a living within any social movement through taking up positions in organizations; the equivalent of writing the history of the environmental movement by focusing solely on those who had jobs working for Greenpeace, Friends of the Earth or the Sierra Club. It would present a picture of the outlines of the movement and would be accurate about the profile, but it would miss the dynamics of the social and cultural change it represents, the breadth of the experience would be lost.

This is not to suggest that farmers and growers are not important within the movement or indeed those employed by the professional organizations in the movement. Rather it means that we need to understand how and why they are important, what role they fulfil and how the movement created that opportunity. These roles stem from the strategy pursued by the movement, which offers people a number of ways of being involved based on their degree of enthusiasm, their situation in life, how much risk they are prepared to take and the context around them. For example the British roads protestors in the early 1990s had to be willing to put themselves at physical risk and live on the protest site, a role that not everyone would be able to or willing to accept (Seel et al 2000). Yet, without other protesters supporting, directly or indirectly, such dramatic protests, they would be futile.

One of the roles that movements give to some activists is the production of symbols to be used by others. The products of organic farms are in many ways symbolic of the aspirations and achievements of the movement. They are both the realization and the ambition of the movement's activities, partially rooted in the present and in part looking forward to the future. We will return to a discussion of organic products, but at this stage it is worth considering the products as symbols and the farms the site of their production. Organic food exists both as a material manifestation of the ideas of the movement but also as exemplars of its ideas, which in part explains why they are often so carefully and critically scrutinized.

Working utopias

The farms themselves act not only as sites on which the techniques of organic farming are trialled and developed, where the physical products grown or raised, but also where the movement has often been socially reproduced. Crossley argues that most movements have 'working utopias' where the goals of the movement are realized, experimented with and debated (Crossley 1999). These allow those involved in the movement to visit or take part in a fragment of the wider goals, to meet others and debate about their shared actions. As they are far more stable and persistent than other manifestations of the movement, they can become places where people gather over long periods of time, in some instances becoming emblematic of the movement as a whole. Therefore farms are central to much of the movement but they are far from the whole story.

Farmers and growers have become, along with some professional organizers, the core staff of the wider movement. The literature of social movements has a great deal to say about Social Movement Organizations (SMOs) and their professionalization (Jordan and Maloney 1997). SMOs are found in most movements providing the resources to coordinate not only campaigns or protests but also to collate information about it for both internal and external consumption. Some are structured like 'organizational weapons', providing a focused tool for conducting particular activities, such as Greenpeace, whilst others facilitate discussion and debate or political lobbying, for example the Sierra Club (Eyerman and Jamison 1989). In turn this can lead to reactions to their tendency to become exclusive, for example, many of the *dis*-organizations of radical environmentalists in the early 1990s were rejecting the example of Greenpeace and Friends of the Earth (Lamb 1996; McKay 1998). In the organic food and farming movement many have come to question the role of farm businesses in relation to the existing large food corporations (see Chapters 6 and 7).

Commerce and the movement

One of the novel features of the organic food and farming movement is that it has a strong relationship with commerce. This is discussed in more detail in later chapters, but it has become the strategy through which the movement has come to public prominence and grown far beyond the bounds it knew previously. The literature on social movements has little to say about the positive relationships between movements and business enterprises. The insights of Crossley do provide a way of seeing the importance of farms within the movement, helping their reproduction and continuity. It does not account for them as businesses.

Tarrow comments that at times movements have adopted the tools of association developed by commercial groups in other periods, and the organic movement has often adopted the organizational form of the farm as an ideal (Tarrow 1996). For some in the movement it represents a form of family based, broadly socially republican enterprise that is the bedrock of a sustainable economy, as can be seen in many writings in the movement.[2] It is also a way of forcing the movement to focus on the practical rather than the aspirational, to focus on what can be done now rather than what might be done tomorrow. At the same time the disposition of contemporary capitalism, which can quickly mimic and simulate products and claims, means that a fuzzy boundary can appear between those businesses that are controlled by the movement and those that are parasitic on it (Thrift 2004).

It is tempting to make a broad division between the organic food and farming movement and the organic food industry or sector, yet such a division would be misleading. It tends to suggest that all the virtue is on one side: that one group is motivated by ideals and the other by money. In application this distinction is hard to maintain, as some people managing large and profitable businesses are more idealistic than those who see themselves at the grass roots of the movement. Rather, it might be more helpful to view this as a spectrum, with some organic businesses deeply embedded in the movement, moving with it and playing an active role in it, to those on the outer reaches of the continuum who have seen a chance to make a profit. However, that such a spectrum exists points to the salience of the movement's strategy of using business and its potential to cut in at least two directions, as will be seen in later chapters.

Mobilization, protest and framing

Although this chapter has emphasized that the organic movement generally does not engage in public protest that does not mean that it has not. The process of moving towards protest is called 'mobilization' in the literature, when a movement gathers its resources to take to the streets or fields. It does not do this in a vacuum; it has available to it a repertoire of protest activities. Some authors have tied this process and the range of protests available to particular cycles of protest, periods when there is an upsurge in conflict and controversy in a society. The events of 1968, the Prague Spring, the riots in Paris, the violent protests at the Democratic Convention, all could be said to be one such cycle, as could the revolutions and protests of 1989. Although participants in the organic movement may have been inspired by, or even directly taken part in these events, the movement has not taken part in such a cycle of protest yet. It has found itself at the centre of a wider wave of protest, which is the subject of Chapter 7.

Generally the form of protest adopted by the organic movement has been a private one, based on the pragmatics of eating and producing organic food, with the symbolic aspects of that as discussed above. When called on to take firmer public action, it has generally taken to the polite and conventional forms of lobbying that many lobby groups have used: letters to legislators, the occasional court action and media campaigns to put the movement's point across. Briefly, at the turn of the millennium, the movement used the forms of protest that its allies were capable of deploying. This was part of a mobilization the organic movement found itself allied with and protected by other groups who have been more concerned with political change and protest than the organic movement. The organic movement has had very little capacity or interest in being able to mount protests.

Social movements are more than the 'how', they are also about ideas, the clash of different perspectives about how life should be conducted. As suggested above, in the discussion of a collective identity and the importance of communities forged by books, at the core of social movements are a set of arguments about how something in society should be. For some authors such as Melucci, but also Eyerman and Jamison, this is the central activity of social movements – to produce and constantly reproduce the cultural codes by which society works (Eyerman and Jamison 1991). The theory is that many social movements are the way in which ideas are put into practice, how new ideas and ways of seeing the world are launched into society. Not only do they produce new ideas, they also construct the 'frames' through which problems are defined. The arguments of social movements realign other arguments to redefine what was previously taken for granted as a problem. In our example, the use of nitrogen fertilizers was uncontroversial until the organic movement 'framed' it as a problem. Equally, it was the framing by the organic movement that made genetically modified plants the particular social problem they have been perceived to be in Europe.

A planetary movement?

The advent of contemporary information communication technologies has seen a flourishing of the idea of global social movements, yet whilst these new technologies have brought innovation and change to movement activity, movements have been global for a very long time. Thomas Paine was able to take part in both the French and American revolutions; the movement against slavery was globalized, as was the suffrage movement (Tilly 2004). People took ideas with them as they travelled, books, letters and pamphlets diffused the ideas and tactics, in a way that the mobile phone and internet do now. The innovation of the new technologies for social movements is the immediacy that they bring and the lessening of the cost of communication. Rather than the weeks required for an exchange of letters, or the days needed for a report to be published, movement activists across the world can communicate with one another almost simultaneously.

As important as the means to diffuse ideas is the way in which a problem and its solutions are framed by the social movement. Some movements are about localized struggles that, as their members become more focused and their thinking becomes more abstract, take on global resonances. Others start with broader horizons, and those concerned with the Earth or the planet tend to have a global frame of reference to start with. In many ways the inheritance of empire that the early organic movement enjoyed was the frame of thinking and acting on the entire world. Issues of the villages and towns of Norfolk in England were seen from the start as the same as those of the Dust Bowl in the US or the veldt of southern Africa. The problems of the soil were those of planet Earth as a whole.

Scholars of global movements, or transnational politics, have identified two important modes of this diffusion. The first is *domestication*, 'the playing out on domestic territory of conflicts that have their origin externally' (Della Porta and Tarrow 2005, p4). Within the European Union this is often where decisions at EU level are translated into protest within the nation state, in part reflecting the democratic deficit of transnational government, suggesting that national governments are still the first resort for many protesters. Chapter 7 discusses the mobilization of the organic movement against GM plants; in part these protests were the domestication of international decisions to a national level. The second trend is the opposite: *externalization*, where the movement targets international or transnational institutions for resources, both material and symbolic. This may in part be driven by the lack of opportunities faced by a movement 'at home'. In Chapter 6, regarding the development of organic standards, this trend can be seen. Long before it was fashionable, the organic movement sought to address the processes of the Earth at a planetary level, diffusing across borders – at first across empires and then through other networks that were bringing people together in novel ways.

Those who walked with Martin

At the inauguration of President Obama, amongst the crowds were those who could talk of having walked and worked with 'Martin' in the civil rights protests of the 1960s and that what they saw before them was in part the realization of their work. As this chapter has shown, social movements are well enough understood to allow their similarities and differences to be compared analytically. So although the organic movement is very different from the civil rights movement in its aims, its context and (we could argue) the extent of its success, there are common features. By following these features, we can gain insights into the development of the movement that should help develop a discussion of the trajectory of the organic movement that has a greater purchase than otherwise would be the case.

The four-part definition of a social movement, that it is made up of informal networks of people who share a set of common beliefs about which they are in conflict with others and that they are prepared to protest about the topics that arise, offers a way of analysing a movement in action. But in a planetary movement, concerned with changing culture rather than legislation or policies, the most direct route is to consider the critical community informing the movement and the discourse around and through which the movement flows. In this chapter I have also set out how we might also identify how the movement spreads across the planet and ensures its continuity through time.

The discussion of the phases of the organic movement in this chapter starts with the opening period of the movement, when it was a loose and diffuse network without the scale or numbers that we would normally associate with the word movement. As the next chapter discusses, the central nodes of that network were to be found in the British Empire and more specifically the UK. These networks developed a critique of agriculture and the dominant form of food production that would launch a global movement. Other insights and ideas from this chapter are developed in later chapters.

Notes

1 www.americanrhetoric.com/speeches/mlkihaveadream.htm includes archive footage of the 'I have a dream speech'.
2 By 'republican' I mean the Anglo-American tradition of citizens who own their own property, run a small business and take part in a democratic process, not the political party of the senior or junior George Bush.

3
Saving the Soil

The ideas, or discourse, of the early organic movement rose during the social and cultural tumult of the period between the two world wars. This was a period that saw numerous features of modern life appear and start to transform many people's lives, leading to powerful reactions for and against these changes. The advent of chemical fertilizers and tractors in agriculture, the motor car, the global transportation of food, movies, airplanes, as well the contest between democracy, communism and fascism saw huge cultural and social changes not just in grand politics but also in daily life. The way that these changes came together in a particular place, Great Britain, created the conditions that allowed the organic movement to begin to form. The UK, as it is now called, was at the time the British Empire, on which – those in favour of it liked to remind people – the sun literally never set. The internal arrangement of this empire gave agriculture a particular role, and in turn those involved with agriculture acquired not only planetary experience but also a unique view of their responsibilities.

The role of agriculture in the British Empire brought together a critical community around questions of food and farming and its consequences for the planet. This community was largely composed of the scientists and physicians who had a role within the empire but also those who, through their political interests, were concerned with the relationship between farming, food and community. During the late 1920s and on through the 1930s, this group were in contact with one another either directly or through their publications. Although far from a movement or even at times a coherent network, they can be fairly described as a critical community. The soil, in the literal sense of its sustainability and fertility, and also the soil as a metaphor for belonging and community, was their problem. Some of their thinking, although arising from the conditions of that time, resonates with the contemporary discourse of the organic movement; whilst some of their ideas are an anathema to the overwhelming majority of those involved with the movement today.

This chapter considers the various actors involved in this critical community. It starts by considering some of the background to the period that had a direct influence on the questions and problems the community was seeking to address. Then it turns to consider a group of physicians who were trying to understand the basis of human health and its relation to food. Next are a group of far-right thinkers and politicians who looked to a connection between the land, racial groups and a new political order. Although their ideology is now defunct, their influence on the movement as it formed later was important and some of their ideas, after being reframed, continue to be important. Finally, the chapter turns to some figures who are perhaps more familiar: Albert Howard and his wives, and Eve Balfour, who is generally known as the founder of the Soil Association. With a rich range of characters in a time marked by extraordinary social and ecological change, this period is central to understanding the origins of the organic movement.

Outside the empire

Outside the critical community discussed in this chapter were two networks that appeared simultaneously but, although significant, have had less influence on the planetary movement. Based in the German-speaking part of continental Europe, these networks drew on an idealism that was quite distinct from the combination of pragmatism and science that characterizes the network in the British Empire. Of these two networks the one founded by Rudolf Steiner – biodynamic agriculture – is the most familiar in the contemporary movement, although the 'Life Reform Movement' that appeared in pre-war Germany set out some important ideas (Vogt 2007).

The Life Reform Movement, like many movements responding to technology and urbanism, sought to get back to the land and achieve a closer union with nature. This task was made difficult by their beliefs that ruled against not only the introduction of artificial fertilizers but also against animal products, including manure, and the use of animals as draught animals as part of their commitment to vegetarianism. The scale and reach of this movement appears unclear but they certainly had a magazine and an organization founded in 1927/8, Arbeitgemeinschaft Naturlicher Landbau und Siedlung, that broadly translates as the 'Natural Farming and Back-to-the-Land Association'. It experimented with organic standards and products sold under a symbol scheme. However, under the Nazi regime, its association was brought into the official government body and the movement did not flourish under the dictatorship. Many of their ideas and publications influenced by their work were, however, published in Switzerland throughout the period. Unconnected with developments in the British Empire, this movement did parallel the attempt to lodge organic farming and food within science.

The largest network and the one with the most direct influence on the development of the early movement was that around Rudolf Steiner. Steiner had already developed an esoteric-magical worldview called anthroposophy and in 1924, a year before his death, he gave a series of lectures about the guidelines that could be used by followers of anthroposophy in relation to farming. These lectures were delivered to about 60 people near Breslau in Silesia but were not published until 1963, circulating for nearly 40 years in the anthroposphic circles in numbered copies, passed from hand to hand. This form of organic agriculture is generally called 'Biodynamic', a term coined by Erhard Bartsch and Ernst Stegemann in 1925 (Vogt 2007). As a shorthand, it is sometimes referred to as 'Steinerism' or 'Steinerist' agriculture, which is a misattribution as it was developed by a network for farmers and gardeners along the guidelines provided by Steiner (Schmitt 2006).

The philosophy underpinning biodynamic agriculture views nature as a mix of the physical and the spiritual intertwined in a matrix that has four levels: the physical, ethereal, astral and ego forces. Whereas many organic advocates may say that their farm is analogous to an organism, for biodynamic farmers the farm *is* an organism that exists on the four levels and can be manipulated on those levels. The farm can and should be a self-supporting system, on which fertility and health are built up, without the need to bring in other supplies. Farmers can manipulate the farm on the astral and ethereal levels with preparations that combine compost and arcane elements. From these esoteric foundations the network of early farmers developed a whole farming system that they had standardized by 1928 under the Demeter scheme. In doing so they put aside some of Steiner's ideas, for example his opposition to green manures, to present a workable system. Hardy Vogt records that from the beginning scientific testing was undertaken to determine whether and if there were any benefits to the biodynamic preparations – with little by way of definitive results (Vogt 2007).

How many organic farms there were in the late 1920s and early 1930s is unclear but estimates range from 100 to 2000, and certainly there were a few large estates that hosted the experimentation and underpinned its development. However large the network was, it was robust enough to develop a range of organizations, publications and experiments, by 1933 forming under an umbrella organization, 'The Association of the Reich for Biodynamic Farming'. They actively promoted biodynamics not only within Germany but also across Europe and later in North America. With its foundations in an esoteric system of belief, it did not constitute a social movement rather a more new age religious one. Self-sufficient, with a dedicated and rather closed network of adherents, biodynamics frequently looked back onto itself rather than to the wider task of transforming the planet's agriculture.

Vogt describes the relationship between biodynamics and the Nazi regime as an 'alliance', arguing that the leadership presented aspects of biodynamics as being compatible with Nazi ideology. Certainly some actively worked with the Nazis, for example, Franz Lippert supervised a biodynamically managed herb plantation at Dachau concentration camp. Senior Nazis such as Richard Walther Darré, Heinrich Himmler and Rudolf Hess took an interest in biodynamics. Yet, as with the paradoxical and often incoherent Nazi regime, in June 1941 Himmler almost simultaneously sanctioned field trials of biodynamic preparations and the suppression of the remaining organizations. The networks of biodynamic agriculture made an accommodation with a dangerous dictatorship, as so many people did then, and individuals within that network responded in a range of ways. Although individual Nazis such as Himmler and Hess were attracted by the esoteric nostrums of biodynamics, it is important to remember where the power lay, and the predicament faced by those forced to deal with such a brutal regime (Bramwell 1985; Vogt 2007).

Food, empire and reform

The cultural and social shock of industrialization and urbanism had started on the other side of the English Channel sometime before it had in Germany. The British were the first to enjoy a global diet thanks to the configuration of their empire. By the beginning of the 20th century the British diet was built around the food produced by the empire. Tea, the staple beverage, was the product of Indian or possibly Kenyan plantations and was likely to be sweetened with sugar grown in the West Indies. The famous roast beef, for those who could afford it, may well have come from the Argentine pampas. For those less fortunate, bread was made of wheat grown on the Canadian prairies, and the jam that topped it again would likely depend on West Indian sugar. This globalized diet had a price, and not least for the colonized subjects. Mike Davis in his book, *Late Victorian Holocausts*, charts how British imperial misrule contributed to millions of Indians starving to death at the turn of the 20th century (Davis 1999, 2002). The imperial diet had less immediate but nonetheless profound consequences for the British working class, as they were the first large group to live on a high sugar, low fibre diet. George Orwell's account of the British diet in *The Road to Wigan Pier* documented the impact that this diet had on the poor of the industrial heartlands of the empire. The circuits of imperial trade left many starving, whilst purveying to others an exotic but often nutritionally inadequate diet.

Imperialism changed not only the colonized but also the colonizers in a number of significant ways (Griffiths and Robin 1997). As the metropole of the empire, the British Isles were supposed to import basic goods and export manufactured ones. Some basic industries such as coal mining needed to be near the factories and industrial heartlands, but others did not. Agriculture was always left in an ambiguous position, as food could be grown more economically in the dominions, which often had significant natural advantages. British agriculture would always struggle to compete against the warmer climates, cheaper land and the 'abundant' labour of other parts of the empire (Perkins 1997). British workers, in this vision, should be engaged in the more profitable and fitting role of manufacturing goods that could be sold back to the rest of the empire. In strategic terms British agriculture was at best often marginal and at worst completely residual to the workings of the realm. It could be left to be the backdrop to the leisured lives of English gentlemen or the sporting estates of Scotland or idealized visions of the mother country (Matless 1998).

The paradox of this position was that whilst agriculture in the British Isles was unimportant, the British Empire was the first to farm across the planet. Within the empire almost every form of agricultural land and farming system was represented from the prairies to paddy fields to mountain pastures to hill top plantations. Whilst the attention of the empire was often on commodity crops – tea, coffee, opium, quinine, cotton and tobacco – that were globally traded, alongside them were a host of other forms of agriculture. The Indian subcontinent had millions of people still following forms of subsistence agriculture that represented the knowledge of thousands of years of growing in those conditions, whilst parts of Australia, New Zealand and Canada had only been farmed within living memory (Grove 1996). Alongside the empire of commerce was an empire of agricultural science. British agricultural scientists were faced with challenges from all over the planet; an individual scientist could expect to travel widely and gain a very varied experience. Although based in British institutions, this perspective gave those involved with it a sense of involvement with natural systems more fragile, far less stable and robust than the temperate example of the British Isles.

The final and most terrible product of the empire was an abiding interest in the concept of race and racial fitness. Throughout the empire were examples of the diversity of human culture that were understood at this time through the lens of race, and it was considered that these were based on biologically distinct branches of humanity. In the struggle for empire, that the British were the pinnacle of supremacy, not just the 'white man', fuelled this interest. Some believed that Britain was an ancient cradle of civilization – as late as the 1920s some held that ancient Egyptian ruins could be linked to southern England (Wright 1996). Others thought that the natural environment shaped the races, part of which was diet, and therefore looked to physical differences in the shape of the face or head as explanations for human variety. This often led to a form of Lamarckism, a belief that the traits developed by an individual could be passed directly on to their off spring. Fitness in these discussions was not only understood in a Darwinian sense, as adaptation to a particular environment, but literally as the bodily health of individuals that could be transmitted between the generations. This manifested itself most violently in the racism shown towards people subject to the various empires of the period, but also in the concept of eugenics – widely popular at this time – and included the less socially advantaged citizens of the empire as well. Any discussion of biology during this period was redolent with metaphors and concepts that went far beyond the idea apparently at hand.

The doctors

Amongst those who were most influential in the networks of what would become the organic movement were a disparate group of physicians who had become concerned about human nutrition and its relationship to the environment. Many of their prescriptions now appear common sense – the virtues of fibre in the diet, fresh fruit and vegetables and exercise, but it must also be remembered that, at that time, tobacco was generally considered benign if not beneficial to health and many of the consequences of dietary choice were unknown. These doctors, dentists and nutritionists forged a new path, that at the time appeared to be cranky support for some of the period's dietary fads.[1] Their concerns raised questions, however, about the relationship between health and diet that deliberately linked to the experiments others were making in farming.

These scientists worked across the globe, in a wide range of communities, and were concerned with questions of well-being and health rather than immediate clinical need. It appears that they were aware of one another's research, even if not in direct correspondence, and long before the books and pamphlets that they used to communicate their ideas to a wider audience. Often they were also able to mobilize the resources necessary to carry out the scientific and social experiments that they thought were required to explore their ideas. It was more than a matter of personal practices and speculation but of exploring alternative models, then disseminating those findings further afield.

The experience of medicine in India exposed British doctors to a wide range of diets and different ways of life. The example of these contrasting diets, through experimentation and how people lived, was a key part of the discussion of diet. Sir Robert McCarrison had published several medical monographs on the thyroid and metabolic disorders, becoming a highly esteemed medic during his career. Whilst working in the Indian Medical Service, McCarrison initiated an experiment into the effects of different diets on health. He fed rats with the various ethnic diets that he saw reflected in the Raj and British society. British working class rats were fed a diet rich in white bread, jam and tea, whilst Pathan rats had a diet rich in vegetables and pulses, and Hindu rats had a diet consisting largely of vegetables. The photographs of the rats showed the 'working class' rats looking rather gaunt, with the 'Pathan' rats looking vigorous and the 'Hindu' ones looking wiry and alert. These findings did not find their way into the scientific literature because of problems with the number of rats used in the experiments. This did not stop the experiments gaining public attention, and McCarrison gave a lecture to the Royal Society in 1936, which was later published as a book (Conford 1988).

The humble dried apricot became a potent symbol of a healthy diet through the work of Dr Guy Wrench. If McCarrison was a mainstream figure, Wrench appears to have been more unconventional. McCarrison offered guidance on nutrition, whilst Wrench linked this firmly with the soil. Hunza apricots owe their name and reputation to their 'discovery' by Robert McCarrison (Wrench 1938, 1939, 1946). The Hunza people lived in the Hindu Kush on the borders of the then British Raj and what is now on the border of Pakistan and Afghanistan. Isolated high in a mountain valley, they had a distinctive way of life and agriculture. Wrench records that at the turn of the 20th century they numbered only 14,000 people, but were noted as porters, agriculturists and artisans. The Hunza were important because they were remote, unworldly or rather an 'Other':

> *Everything suggests that in its remoteness it may preserve from the distant past, things that the modern world has forgotten and does not any longer understand. And amongst those things are perfect physique and health.* (Wrench 1972, p22).

Living on carefully gardened and farmed terraces cut into the mountainside, with a diet of fresh vegetables, sprouted seeds, a little dairy produce and meat, coupled with physical exercise, they acted as exemplars of health. As a people, the Hunza represented both the past and a possible future, in that they offered the possibility of recovery. Wrench's work on the Hunza has been reprinted regularly, most recently in 2006, as it stands as an accessible example of the rewards of a healthy diet.

The Peckham Experiment

From the distant mountains of the Hindu Kush to the heart of the metropolis, 'The Pioneer Health Centre' or the 'Peckham Experiment' in London was the leading institution for a group of health practitioners concerned with the conditions that fostered health. It represented another possibility for the creation of a socialized health service, which encompassed the social, familial and environmental health of individuals (Pearse and Crocker 1943; Scott Williamson and Pearse 1947; Pearse 1979). For those behind the Peckham Experiment, the scale of the project was determined by the needs of health:

> *for experience had already taught that health could only come forth from mutuality of action within a society sufficiently mixed and varied to provide for the needs of mind and spirit as well as of body* (Pearse and Crocker 1943, p6).

The Pioneer Health Centre was an experiment conducted to discover the 'nature of health'. It ran in various forms for nearly 20 years and created a group who were amongst the most innovative and creative activists in the organic movement. Ironically, it was the creation of a National Health Service in 1948 that removed the space for such an experiment. Innes Pearse worked as a physician in several London hospitals and became the first female registrar in a London hospital. George Scott Williamson, her fiancé, was a pathologist in the London teaching hospitals. Their personal partnership extended into their professional lives as they created social experiments in the search for the 'ethology' of the human species.

Starting work from a terraced house in 1925, they started to survey the health of the local population. After four years, they closed the project and spent the next six years planning the next phase. The Pioneer Health Centre opened in a purpose built facility in 1935. It was intended as a laboratory for exploring the autonomous behaviour of humans, to discover the true character of health. For Pearse and Scott Williamson, the true purpose of health was to realize the full potential of the human being. This potential could be found through the vehicle of the family in full and free association within a diversity of other people. The Centre provided a swimming pool, a gym, a self-service restaurant and other leisure facilities. For a small weekly sum, local families could attend the centre, in return for having regular physical examinations or 'overhauls'.

To discern the ethology of humans, the centre had no organized events or structures although participants were free to organize their own; the only rules were that no one person or group should come to dominate others. A staff member would arrange for the allocation of facilities but no more. Even the cafeteria was self-service; the first in the UK, to ensure that people ate what they chose:

> *a healthy individual does not like to be waited on; he prefers the freedom of independent actions which accompanies circumstances so arranged that he can do for himself what he wants to do as and when he wants to do it* (Pearse and Crocker 1943, p75).

The role of the scientists was to observe and to assist the members of the Centre. After the physical 'overhauls', the results would be discussed with the family and the possibility of remedial action discussed (Aitchtey 1995). Other physicians would provide treatment or the centre itself would suggest an adjustment to diet. The scientists themselves were immersed in the structure of the Centre, their objective being to help others achieve their potential for health through the most democratic means possible: 'The Centre has in fact, shown itself to be a potent mechanism for the "dem-ocratisation" of knowledge and of action' (Pearse and Crocker 1943, p78).

Appointments and arrangements were made to encourage the autonomy of the members who, it was thought, had had their opportunities for self-expression suppressed by the disciplines of industry. Physical 'overhauls' produced what for those involved was the shock result that only 9 per cent of participants were healthy (free of disease or disorder) on their first visit. It was considered that for many of the adults attending the centre, the ravages of poor health and chronic illness damaged their ability to realize their full potential. The next generation presented the opportunity for full health. Such health started before conception, which ideally would be planned through advice available from the Centre, and the new life would be nurtured from that point. The Centre's own 'Home Farm' provided 'organically' grown vegetables and fruit, along with milk tested for tuberculosis.[2] Priority of supply was given to expectant mothers and children. From the reports of the Centre, it is apparent that an analogy was being made between the health of the soil and that of Humans:

> *Tilling the familial and social soil of man is becoming a science and art to be acquired with all the assiduity – and more – given by man to the study of physical phenomena and to the study and cultivation of his plants and beasts ... Health is a cultivator's problem, and that cultivator can ultimately be no other than the biologist* (Pearse and Crocker 1943, p123).

To continue with gardening metaphors, the ground for future collaboration had been prepared.

Within the Centre, the focus of research was placed on the family. No one could join unless their whole family did; great emphasis was placed on what they described as the 'mate pair', although this initial stricture was stretched to include couples without children, widows and widowers. They were socially innovative, offering family planning advice to couples from within the Centre. Certainly, in their discussions, the founders of the experiment went to great lengths to ensure the moral probity of their actions. The 'truth' they were seeking was to be able to redefine the human organism: 'In the functional sphere man and woman do not work reciprocally as in mechanism, but mutually as diverse parts or organs of a unified organism – like a small ant heap linked in the continuity or what later we shall have to call the "specificity" – of a "functional organisation"' (Pearse and Crocker 1943, p18). Only as a couple could humans find the health and wholeness which would make them truly alive. The couple could only find their fullness in the company of a broader society; later the metaphor of a hive was adopted: 'And so with the Centre – it merely gives the frame; the family secretes the specific honey, each family contributing to society its own peculiar flavour and quality' (Pearse and Crocker 1943, p131). This was not to be a society of drones ruled over by soldiers or aristocrats, but a free association of diverse individuals. Wartime evacuation and conscription left the Centre without the very families it needed. The building was requisitioned for the duration of the war, the swimming pool rather symbolically being filled in with concrete, and the whole building used as an armaments factory. Although the war appeared to end the experiment, in a curious manner it served to bring those involved with the project into contact with a wider group of people.

Later chapters discuss in detail the role that the staff of the Peckham Experiment played in the formation of the Soil Association. Not dependent on farming for a living, they demonstrated the ability to innovate and administrate informed by their experience at the Pioneer Health Centre. In contrast to the other medical elements of the critical community, the Peckham Experiment was a scientific social experiment. Behind it was a notion that people could gather and organize themselves without guidance. It was carried out beyond the auspices or intervention of the state; rather it relied on the voluntary association of individuals. At a time when many saw that state as being central to all political and social projects, this was highly unusual. Implicitly it saw a link between human health or well-being and the expression of freedom. Whilst others responded to the crisis of democracy of the 1930s with various forms of authoritarianism, of either the Leader or the Vanguard party, the Peckham Experiment invested in the self-service restaurant. The crisis of democracy could only be solved by more democracy – not less.

Agricultural scientists

In the late 19th and early 20th centuries several highly influential books were written about farming in Asia that caused some Western agriculturalists to think differently about their ways of farming (King 1927). A starting point is always somewhat arbitrary, but it would seem that the terrible famines in British India would suggest a good point to begin. Famines in the Raj appeared with terrifying frequency, imperial tax demands and ecological fragility meant that the people of the subcontinent were very vulnerable and the imperial rulers were not immune to the suffering that they caused, not least because of the loss of tax revenue (Davis 1999). After the famines of the early 1880s, the government's famine commission recommended considerable agriculture improvement. A report into the possibility of such improvement was written by John Augustus Voelcker and published in 1883 (Voelcker 1883). Voelcker was the 'Consulting chemist to the Royal Agricultural Society' and therefore considered to be well qualified to pronounce on Indian agriculture. Voelcker's report became the urtext of organic agriculture, in finding the traditional agriculture to be excellent:

> I, at least, have never seen a more perfect picture of careful cultivation, combined with hard labour, perseverance, and fertility of resource than I have seen at many of the halting places on my tour. (Voelcker 1883, p11)

Indian agriculture was not the problem, just the bad practice of it in some areas and the lack of facilities more generally. Although using the Victorian terminology of 'race', Voelcker was more concerned with the inequities and inefficiency of the caste system. What were required first were not Western methods but more good Indian practices spread more evenly; he had no doubt that this could be achieved:

> *The Native, though he may be slow in taking up an improvement, will not hesitate to adopt it if he is convinced that it constitutes a better plan, and one to his advantage.* (Voelcker 1883, p11)

Implicitly, the report was a critique of the social arrangements of Indian agriculture not their technical efficiency, a social critique launched by an imperial servant and one who might have been expected to be the proponent of the latest technologies in farming. As Voelcker travelled out to India, he met Robert H. Elliot (Voelcker 1883, p7) who would later popularize growing permanent pastures and lobby against artificial fertilizers. Voelcker comments of Elliot: 'besides as an able writer on Indian agricultural matters. From him I learnt much that was afterwards invaluable to me' (Voelcker 1883, p7). Indeed, such was their friendship that 20 years later, Elliot records Voelcker's visit to his farm in 1904 (Conford 1988, p111). The experience of Indian agriculture, a critique of Imperialism and proto-organic agriculture were beginning to become intertwined.

Neither Voelcker's report nor his enthusiasm for Indian agriculture could prevent famine striking again, as it did in India between 1898 and 1902. After feeling confident that famine could not return again on the previous scale, the Raj administration behaved with a ruthlessness that appalled journalists reporting to the London newspapers. Millions perished as disease and plague, compounded by miserly rations, squalid camps and forced migrations, were put into harness with the drought and, as a consequence of this: '[in] many parts of India there had been a fifty year standstill in population growth' (Davis 2002, p175). The new Viceroy, Lord Curzon, sought to move imperial India out of this horror through, again, agricultural improvement; part of which was to attempt to improve Indian agriculture through education and the appliance of science. Although he had a number of grandiose plans, most of which were not approved by Whitehall, Curzon managed through a private donation to set up the Imperial Agricultural Research Institute (IARI) at Pusa, in Bengal. To run the project he appointed Albert Howard as the first Imperial Economic Botanist.

Albert Howard had trained at Cambridge and had worked previously in the Caribbean and at Wye College, near London, before taking up this post. With him came his first wife, Gabrielle, who had also completed postgraduate studies at Cambridge. The two worked as a team throughout their career in India, bringing with them both the latest knowledge of the application of genetics to plant breeding, and questioning minds. Not only was the IARI the only such institute in India, its like did not exist in Britain. As John Perkins notes: 'It is ironic but telling that the British Government fostered an invigoration of agricultural science in India before it turned to its own land and agricultural economy' (Perkins 1997, p81). Quickly, the Howards became leading imperial wheat breeders, with a stream of books and papers to their credit, as well as new seed varieties. Later in their careers, they began to deviate from the path of Imperial servants ascribed for them. Gabrielle started to formulate new ways of considering the role of health, of both plants and of humans. Research into the role of air in the soil changed its status for the Howards and simultaneously they started to question just whom they served. Like Voelcker before them, they began to view the practice of the Indian cultivators around them positively. Their mission became one of explaining traditional agriculture in scientific terms and in so doing improving its rigour. During the 1920s, the Howards shifted decisively, leaving the IARI for a new research centre at Indore, which they founded with private sponsorship. Increasingly, they saw a triumvirate appearing, with the health of the soil being linked to that of plants and through them to the health of the animals which ate them. Artificial fertilizers destroyed the soil, reducing its ability to resist disease and so on, along the chain of health (Howard, A. 1940, 1945; Howard, Y L. 1953; Geiryn 1999; Barton 2001). During the 1920s, the Howards reinvented the Hippocratic tradition and applied it to agriculture.

Their central and transforming technological innovation was composting, carried out through what they named 'the Indore method'. Compost was a redemptive technology, and one that pre-figured calls for more appropriate or intermediate technology (see Chapter 5). Compost was designed to return not just nutrients to the soil but also humus and to improve the soil structure. This developed the work of the Howards during the 1920s regarding the role of air in soil. It also started to become the resolution to a multitude of problems, from soil erosion to the lack of minerals in food crops. For those concerned about the soil as an emblem of their contemporary ecological crisis, compost – an artificial soil supplement made through the emulation of natural processes – offered redemption. According to the implicit imperial service tenor of the Howards' work, it was quite literally freely and easily available, whilst artificials required huge processing plants and chemical technicians (Perkins 1997). For the Howards, science was about service to the community, not the self, and they became fierce opponents of the use of science for personal advancement (see Chapter 7). Compost required very little equipment. Using waste materials such as manure, peelings, straw, cast-offs and then dousing them with urine, the compost could be put in a box or container, or it could be just heaped. A brief leaflet or simple demonstration could make it available to anyone and everyone, for nothing but the effort of gathering the materials. Compost was low tech and low cost, ready for use within months, and it could arrest, even reverse, the gathering ecological crisis.

After Gabrielle's death in 1931, Albert Howard retired from imperial service and returned to Britian. He became part of the critical community, championing his insights and denigrating the role of 'artificials'. During this period, the Howards' redemptive technology started to cause a split in the critical community between the biodynamic farmers, followers of Rudolf Steiner and those who were not. Albert Howard was one of the first to use, if not the contriver of, the locution 'muck and mystery' about biodynamic farming. At a time when the works of Steiner were new and many experimented with a variety of techniques, Howard started to draw rigid lines. At this time most of those in the critical community were experimenting with ideas drawn from those around Steiner as well as their own ideas. Albert Howard, with his insistence on scientific method, offered another path, separate to that of biodynamic farming. Where Steiner had posited hermetic terminology and neo-magical formulations, Howard offered observation and field science more in tune with the dominant ideas of the time.

Using Albert Howard's formulation, the advocates of 'organic' agriculture did not have to reject science totally but embrace a different variant of if it. Howard led from observation: 'The plant or the animal will answer most queries about its needs if the question is properly posed and if its response is carefully studied' (Howard 1940, p59). In place of gnostic ruminations about the whole, he offered a cyclical model: 'All phases of the life cycle are closely connected; all are integrated to Nature's activity; all are equally important; none can be omitted' (Howard 1940, p23). The fundamental cycle of all of these for humans was to be found in agriculture, but Howard expressed this in a very particular way: 'The efficiency of the green leaf is therefore of supreme importance; on it depends the food supply of this planet, our well being, and our activities' (Howard 1940, p23).

Given the importance of science for Albert, as for his late wife, agricultural science was an ecological pursuit. It was also a public service and 'complete freedom' was necessary for researchers, from both the state and private companies. Howard did not fight shy of making social pronouncements related to his interests, especially in the early years of the Second World War. He was determined that food should be of paramount importance: 'The nation's food in the nature of things must take the first place. The financial system, after all, is but a secondary matter' (Howard 1940, p198). He was also able to hint at how some of the social relations might be arranged: 'The situation can only be saved by the community as a whole. The first step is to convince it of the danger and to show the road out of this impasse' (Howard 1940, p220). It is unusual in the language of rural reform of this period to see the term 'community' used. Howard continues reinforcing this with a call for fraternity: 'All engaged on the land must be brother cultivators together' (Howard 1940, p222). Finally, he had suggestions about how to ensure that humus grown food won through: 'Foodstuffs will have to be graded, marketed, and retailed according to the way the soil is manured' (Howard 1940, p221). Howard offered a way through organic agriculture that did not abandon science, although it embraced a version of science that would be marginalized during the green revolution.

Albert Howard, although a radical in his own times, did not stand outside them. Like Wrench and Balfour he viewed the Earth as a feminine actor, capable of concerted action: 'Mother Earth, rather than the advocates of these various views, will in due course deliver her verdict' (Howard 1940, p59). Although he considered Indian peasant farmers his 'brothers', his fraternal feelings did not extend to all of humanity; he suggested some experiments about nutrition should be confined to a particular group: 'The only subjects that might conceivably be used for nutrition experiments on conventional lines are to be found in concentration camps, in convict prisons, and in asylums' (Howard 1940, p171). Although Howard may have drawn numerous innovations from the imperial experience, his imagination also encompassed the archipelago of experiments which could be found within the empire of science then. He straddled the position of having rejected the claims of Imperialism without rejecting all of its machinery.

The agrarian far right

Through the 1930s there was a complex constellation of groups and parties involved in the politics of the far right – a spectrum that stretched from ultra-conservatives to nationalists to fascists and national socialists. The ultra-conservatives looked backward to the revival of the role of the aristocracy and the monarchy in the political life of the empire, whilst the national socialists and fascists looked to a modern society dominated by the state through the personage of the great leader (Webber 1986; Baker 1996; Griffiths 1998; Thurlow 2006). These groups often shared little in the way of political outlook other than a fervent opposition to socialism or communism and an enthusiasm for the continental dictatorships of Mussolini, Franco and later Hitler. As is often the case with marginal groups, much of their energy was spent in factionalism that was opaque to those outside those groups. Two of these groups had a particular significance for the early organic movement. It is perhaps easiest to track the ebb and flow of these groups by following the careers of particular people.

Rolf Gardiner remains one of the more significant characters of a small network of romantic far-right-wing activists who became part of the early organic movement. Gardiner was from an Anglo-German background, and spent the immediate period after the First World War reviving English folk dancing and exploring mysticism as well as the German youth movement. Throughout his life there was a link to developments in Germany: during the 1920s youth movement and then aspects of National Socialism.[3] Gardiner was a cosmopolitan figure with estates in Africa, as well as properties in Europe, but rural Dorset became the base of his activities (Best 1972; Chase 1992; Wright 1996; Moore-Colyer 2001a, 2001b).

In the late 1920s Gardiner bought and renovated the Springhead Estate, in Dorset. It was as an idea he had clearly explored in correspondence with the author and poet D. H. Lawrence – who imagined it in this way:

> We'll have to establish some spot on earth, that will be the fissure into the underworld, like the oracle at Delphos, where one can always come to. I will try to do it myself. I will try to come to England and make a place – some quiet house in the country – where one can begin – and from which the hiker, maybe, can branch out. Some place with a big barn and a bit of land – if one has enough money. Don't you think that is what it needs? And then one must set out and learn a deep discipline – and learn dances from all the world, and take whatsoever we can make into our own. And learn music the same; mass music, and canons, and wordless music like the Indians have. And try – keep on trying. It's a thing one has to feel one's way into. And perhaps work a small farm at the same time, to make the living cheap. It's what I want to do. Only I shrink from beginning (D. H. Lawrence to Rolf Gardiner December 1926 (Aldington 1976, p16)).

Gardiner did not shrink from beginning; he made Springhead a centre of music, singing and dancing, as well as hosting the camps of the English Mistery and later the English Array during the 1930s (Stone 2003).[4] Here he brought people together in a running cultural experiment, as well as one in agriculture and forestry.

Gardiner's influence grew through his membership of the English Mistery. The Mistery was working towards a revival of British greatness, which they felt had been undermined by the Irish home rule Acts, the rise of the labour movement, the extension of democracy and the rise of the feminist movement. The programme of the Mistery, such as it was ever publicly expounded, involved the restoration of the monarchy, the assertion of manly values, the development of leadership skills and the revival of rural Britain. What they had in mind was an organic society; the overwhelming metaphor they used was that the nation was a body, a biological entity, in which everyone had a function. Hence outsiders were suckers, parasites. The Mistery was open in its general racism and more specifically its anti-Semitism.

The founding book of the English Mistery was *Statecraft* by William Sanderson in which he revealed that the 'ancient secrets' had been lost, and unless they were restored the English as a 'race' would continue to degenerate (Sanderson 1929, 1933). Only through the re-establishment of a landed aristocracy could it be ensured that the race would continue and thrive. The Mistery dedicated itself to the recovery of these secrets through 'Leadership'. Hence, the Mistery was broken down into subgroups of 'Kins', which were led by an individual. No votes were ever taken and the leader's opinions were paramount, which was a facet of the Mistery's more general contempt for democracy, although this did not prevent its members from standing for elected office. Leaders were supposed to demonstrate aristocratic virtues, a *noblesse oblige* that was to be reciprocated by those grateful for being ruled so wisely (Sanderson 1929).

Sanderson listed the seven lost secrets as: Memory, Race, Government, Power, Economics, Property and Service. As arcane knowledge, they were revealed only to those in the group. Race was the entry point for understanding all of the secrets, with race being a eugenic and Lamarckian concept (Sanderson 1933). Memory was to be preserved through race, rather than the abstractions of learning. Government was to be exercised by a new aristocracy based on race, eugenic superiority and induction into this memory. Power in turn was to be held by this aristocracy, with the monarch at the zenith, but a strong element of *noblesse oblige* was observed. Property and economics were similarly tied together, with private property giving individuals the opportunity to inherit and so ensuring their responsible stewardship of it. Economics had been largely lost to moneyed interests, which when combined with the secrets of race, meant Jewish interests (Ludovici 1935; Stone 1999). Members were initiated into the secrets through quasi-Masonic rituals, and the Kins appear to have been formed of influential upper-class men.[5]

Although Sanderson had conceived of the ideas underpinning the Mistery, he did not lead it – that role was taken up by Gerald Wallop, later Lord Portsmouth. Portsmouth was from an Anglo-American background, had fought in the First World War, served in Parliament and was thoroughly dismissive of democracy. As Portsmouth admitted later in his life, the Mistery 'coloured all my political thought after late 1930 and dictated most of my standards of value' (Portsmouth 1965, p126). Wallop was both an interpreter of these ideas, and a propagandist of some skill. His books were influential beyond the narrow circles of the far right that he mostly operated within. *Famine in England* in 1938 set out the importance of agriculture and a vibrant rural life, particularly for a nation likely to be facing war in the near future. It provided a statement of the importance of agriculture; not only for the production of food but also that it was the breeding ground of the English (Conford 1988). It was a statement of the superiority in racial and cultural terms of the rural life and rural people. Wallop saw cities as the breeding grounds of 'half-breeds' and the wellspring of 'godless' communism (Portsmouth 1941). It was this book that inspired Eve Balfour to take an interest in the relationship between soil and health (Conford 2001).

Wallop campaigned for the appeasement of the Axis powers, as well as against the steel plough and asylum for Jewish refugees from Nazi Germany. His political activism brought him close to many in the Fascist groups and movements, although his sense of patriotism appears to have prevailed (Griffiths 1998). As the 1930s grew darker, the Mistery was wound up by Wallop and Sanderson was expelled; it was then re-formed under Wallop's sole leadership as the English Array. The Array was far more sinister than the Mistery with, as the etymology of the name suggests, militaristic undertones.[6] Dan Stone comments: 'This congregation of Fascists made the Array serious in a way, which the Mistery had never been' (Stone 1999, p205). Integral to all of this for Wallop was a vibrant and restored agriculture, from which would flow the aristocratic order and revived race he saw as so essential. Simultaneously sinister and silly, these ideas and those who propagated them, were to play an important role in this critical community and the early years of the organic movement.

Eve Balfour

The person most associated with the founding of the organic movement, and the Soil Association in particular, is Lady Eve Balfour. Although undoubtedly a remarkable individual, she also occupied a number of positions that made her ideally suited to being a figurehead for the movement. Balfour was from a profoundly aristocratic background, the niece of a former prime minister, which made her an insider in a class bound Britain. Her background and her privileges also spared her from the demands of having to have a conventional career. Yet as someone who embraced modernity and lived a decidedly bohemian life, she could also reach past the narrow confines of establishment. Balfour wrote detective novels, played in a jazz band and learnt to fly, she was far more involved with, and embracing of, the modern age than others in the critical community. Younger than many of the other members of this dispersed network of thinkers, she also lacked some of the experiences that had so marked their lives. Balfour was the bridge between the critical community and the movement that grew up around it (Brander 2003).

Balfour's particular contribution to the critical community was that she was an organizer and a synthesizer. Her talent for bringing people together around specific projects meant that she was often at the centre of activities. Although through her various efforts she helped the movement make progress, the intellectual breakthroughs were made by others. For some commentators, such as Patrick Wright, her role was to purge the organic discourse of its fascist and racist elements: 'We can only marvel at the fortitude with which Lady Evelyn Balfour laundered Lymington's [Walbp] books to produce the cleaned-up ecological vision on which she founded the Soil Association' (Wright 1996, p175). Balfour's biographer Michael Brander quotes a letter from this period, as she thought about how to nullify the influence of Mosley's British Union of Fascists in rural areas (Twinch 2001).

> *All of these views are held and preached by and people forget his ghastly methods, his anti-Semitism, his intolerance. I want to see a new party emerge that can legislate with the same end in view, but without dictatorship ...* (Brander 2003, p37)

This was in many ways mainstream conservative thinking in the period, although Balfour does appear to be particularly alive to the racism about which many others were very casual.

It was at the end of the 1930s that Balfour decided to conduct an experiment to investigate whether organic farming was superior to other forms of agriculture. Although there were small-scale efforts to farm with compost, or to avoid artificial fertilizers, no one had conducted a side-by-side experiment, with an organic unit run alongside a non-organic one. It was this that Balfour set about organizing as war broke out across Europe. Her farm at Haughley in Suffolk would become the place where the ideas of those critical of artificial fertilizers might be validated. Balfour demonstrated her talent for creating organizations by placing her farm and that of her friend and neighbour Alice Debenham into the Haughley Research Trust, with the explicit aim of discovering if the condition of the soil could influence human health. Over the 236 acres of the farm, she and those involved in the Trust would attempt to prove a very complex and ambitious scientific question. As the new Trust entered the war, it was on a firm organizational basis if not a financial one (Balfour 1943, 1975).

As the decade came to a close, many who had been forecasting a war for most of the 1930s saw it come to pass, as Europe was finally plunged into a conflict that would later draw in most of the rest of the world. The war ended most of the experiments of the 1930s and saw the political activity of many of the groups curtailed, most symbolically at the Pioneer Health Centre (Scott Williamson and Pearse 1947). The Haughley Experiment managed to gain exemption from the rigours of the need to maximize food production and was allowed to continue. As we shall see in the next chapter, the war presented an opportunity for the critical community, as much of what had taken place before the war was recorded and diffused.

The private experiments that were in some ways the hallmark of this critical community also made it very difficult for a larger group to gather and organize around them. Their rural locations and global reach meant that it was hard for large numbers of people to visit them, let alone on a regular basis. There seems to have been a lively circuit of talks at the Farmers' Club in London, and this certainly appears to have been a venue for elite gatherings, but it was hardly the groundwork for a mass movement (FitzGerald 1968). Equally, by this point, many of these people were in the middle if not towards the end of their lives.

Within the discourse of this network was a great deal that those in the contemporary movement would have found attractive: the emphasis on the ecological limits to agriculture, the importance of good, fresh food to health and the links between healthy rural communities and 'organic' farming. At that same time there were those who thought that experiments on concentration camp inmates might be acceptable, that Jews were conspiring to control the world's economy, that women were biologically inferior and that democracy was not as attractive as aristocratic dictatorship. Elements of this discourse would continue, not least in two important strands. First, although conservatives in the broadest sense they were also utopians, they had a vision of a new society and were prepared to be scorned or ridiculed as they attempted to realize it. Second, they believed the world was on the edge of imminent collapse, that the ecological underpinnings of 'civilization' were eroding and this required radical, urgent action. Natural limits were either being met or were about to bite deeply; in their case this referred generally to the soil, but also to the rapid decline of the race (Coupland 1998; Pepper 2005). This combination of utopianism and limits reappear frequently in the ideas that pushed organic food and farming forward in the following years.

This critical community communicated largely through pamphlets, letters, speeches and the occasional visit to one another. Although concerned with the same issues, working away at the same problem, they did not yet share any common organizations, goals or norms. As with many other small and marginal groupings, a fair degree of their energies were taken up disagreeing with one another. Yet their books were read and speeches attended by a wider audience; through describing not only a problem but also the possible solution, this diffuse collection of networks was laying the groundwork for a future movement. Arguably though, the differences between them would have been too great to allow the emergence of solidarity or a shared platform without the intervention of the Second World War.

The war brought about new opportunities for the critical community; food took on a new importance. Shortages controlled through rationing were experienced throughout Europe, whilst starvation was known again in parts of India and China, as food supplies failed totally. This experience began to reshape how food was thought about, how the strategic role of agriculture was viewed and what health meant in a world transformed. It also led to many of the major ideological differences between the critical community collapsing or at least being relegated. Also, more practically, it led to many members of this community communicating directly with one another, forming new personal alliances and writing up their experiments and experiences of the preceding decade. In the next chapter I describe how the first stirrings of a movement and its organization took root in the post-war period.

This diverse and difficult critical community, in a few years of experimentation and discussion, set in place a constellation of ideas ranging from the magical to the scientific, from the democratic to the dictatorial, that proved to be so resonant and productive that some of them are still valued by the movement today. They were not trying to found a planetary movement but to address the urgent ecological and agricultural problems of their day, with the resources they had at their disposal. Although this community was based in Great Britain it was echoed by a similar development in Germany that did not have the same opportunity to flourish. The problem of how to sustain humans and create a healthy agriculture had been engaged with in a manner not seen before; the struggle would become how to find and realize the solutions this grouping had identified.

Notes

1   With very little dietary knowledge, there were some very unusual diets and exercise routines, many of which were followed by those in the organic movement. Sir Stafford Cripps, during the 1930s a member of Parliament for the Labour Party and the first post-war Chancellor of the Exchequer, followed a 'sunshine diet'. On a visit to Washington DC in the 1940s, the ambassador was alarmed to see him jogging naked around the grounds of the embassy.

2   It seems that the farm was managed along biodynamic lines at this time.

3   Youth movements, such as the scouts and woodcraft folk, were part of the response to the end of the First World War. In Germany they became part of the politicized and conflictual situation of the Weimar Republic.

4   Many of those who sang and danced in the Springhead Circle through to the 1960s seem to have been unaware of Gardiner's politics.

5   One member of the Kinship was a minister in the British government.

6   Array – an archaic British term for an arrangement of troops.

4

Poisonous Elixirs

This chapter covers a long period between 1939 and the early 1960s when the organic movement gained a new purchase on events. It is a time that covers perhaps the organic movement's darkest hours through to what many would consider to be the new era of organic farming and food. Its bleakest moment was when, in 1940, a proposal was made that an expanding empire should turn all of its agriculture over to organic methods. The proposer was Walther Darré, the Nazi minister for agriculture, and Hitler rejected his suggestion, although that did little to dampen Darré's enthusiasm. In the immediate post-war period, the by then indicted war criminal Darré was busy trying to set up an organization to educate the German people about organic agriculture, similar to the Soil Association just founded in Great Britain (Bramwell 1985). Darré had heard about this development through correspondence with people well placed in the new organization who continued to share at least some of his political beliefs. An equally symbolic ending point for this period was the spreading of chicken manure over a trial plot at a research station bringing to an end 30 years of experimentation and a phase of the organic movement.

Whilst much of that period has been left behind, at the same time many of the features that we now identify with the movement – the rejection of synthetic pesticides, the questioning of the intensification of agriculture and the certification of goods as 'organic' – were formulated. Many of the key organizations, the Soil Association and the Rodale Institute, that now coordinate the movement, were created and took their first, difficult decisions, as the first dreams of a global organization were outlined. Out of the maelstrom of the Second World War the organic movement was transformed and reinvigorated, now as a transnational movement in a way that it was not before. Yet at the same time it was very marginal, far from a mass movement but one that was tenacious and determined, far from defeated but largely outside the public debates until the intervention of Rachel Carson.

This chapter covers the period that saw the founding of the Soil Association in Britain based on the experience of war and the surge of books about the experience of the 1930s. It then discussed the formation of the Rodale Institute and the changing context of post-war agriculture, in which scientific proof took on a new form. From this point it considers the attempts to set up a global organization to coordinate the global organic movement, an impetus that finally found its form in the early 1970s. It then turns to consider how the discourse of the movement began to turn from the qualities of the soil to take on a critique of the use of the new pesticides and the way in which that shift created new opportunities for a movement that at times had seemed to be becoming increasingly marginal.

The war years

War is often associated with radical and rapid social change. For the organic movement the war brought a number of significant changes, the opportunity for some to write about their experiments in the previous decade and the chance to travel the planet. It also accelerated the technological changes in the use of pesticides and plant breeding that were to underpin what became known as the green revolution that transformed post-war agriculture.

As outlined in the previous chapter, amongst those involved with the movement at this time were some people not unsympathetic to the Axis powers. Some in the broadest networks of the organic movement found themselves collaborating with or actively supporting the Axis powers, others fought and died in opposing them. One biodynamic gardener planted a biodynamic garden at the concentration camp at Dachau, whilst in Britain Jorian Jenks was imprisoned because of his Nazi sympathies. Others, such as Adolf Riechwein, who had taken part in Rolf Gardiner's summer camps in Dorset, were involved in the July plot against Hitler and were executed by the Gestapo (Reed 2004). Meanwhile Reginald Dorman-Smith of the Kinship in Husbandry became minister for agriculture, and during his short spell in government oversaw the 'Dig for Victory' campaign that encouraged the British to grow their own vegetables.[1] He then served as the British governor of Burma during the Japanese invasion. As Roger Griffiths observed in his study of these far-right groups, for many patriotism trumped ideology during these years (Griffiths 1998).

For many who had been active in the pre-war organic movement, the war, paradoxically, proved to be a time of inactivity; they were too old to serve in the war effort yet unable to travel and associate in the way that thay had previously. This was most marked in the British organic movement, which produced a collection of books written and published during the war that recorded both their pre-war experiments and their hopes for the post-war period. Although large gatherings or public meetings were more difficult in wartime Britain, private meetings were possible and many people met during this period who had not previously.

The Kinship in Husbandry

Portsmouth considered himself to have been fortunate not to have been interned because of his presumed sympathies towards the Axis powers (Portsmouth 1965). It was Gardiner who took the lead in keeping the pre-war networks active through a loose network called the 'Kinship in Husbandry'. Although it was not a secret grouping, it was not the public groups that had typified the pre-war period. As part of it Gardiner had recruited people sympathetic to a 'rural revival'. He recruited poets such as Edmund Blunden, journalists such as H. J. Massingham, the historian Arthur Bryant and the landowner Lord Northbourne.[2] Gardiner appears to have viewed it as a semi-secret group that would aim to influence post-war Britain through 'percolating' other groups and introducing their agenda. Others in the group, such as the historian Arthur Bryant, wanted it to be a public organization and despaired at Gardiner's style of managing the group, as well as some of his aims. Gardiner's goal that members of the Kinship would percolate into other organizations and guide them was realized.

Most of those in the Kinship were deeply conservative in their outlooks and political programmes. As Bryant argued in the papers of the group, 'a society can be soured by being stirred too deep' (Bryant 1943). He continued:

> *Ideas now at a discount may not be so in 20 or even 10 years' time, but mean-while we must be ready to seize opportunities. It is no use fighting against the stream of politics – the Fabians saw that – but we can permeate existing organisations, probably best working as individuals in our spheres.* (Bryant 1943).

Some who had been invited into the Kinship quickly left; the journalist Adrian Bell[3] was so disturbed by Gardiner's ambivalence towards the war that he left in order to be disassociated from him (Gander 2001). Members of the Kinship did appear, without declaring their allegiance, in other organizations, as will be discussed shortly.

The most important public product of the Kinship was a book published in 1945; a collection of essays, edited by Massingham, by authors all of whom, bar one, were part of the Kinship. In the collection, Portsmouth proposes a system of local food based on a rejuvenation of the great estates, where the landowner would coordinate the production and distribution of food locally. Gardiner, in his chapter, praised the use of forest products by the German army, as an example of the importance of rural resources. Massingham, in the introduction, criticized the Nazis for too much blood and not enough soil, of having broken with rural traditions and embraced industrialism. He saw in the beginnings of the British welfare state that: 'Our people are not thinking in terms of self-help responsibility, but of social services and security which can only mean the authoritarian state' (Massingham 1944, p5). In this work the Kinship demonstrated that their conservative agrarian ideas and values had survived the war and they were hoping to use them to shape post-war society.

> **Box 4.1** *Rolf Gardiner's War*
>
> Rolf Gardiner suffered from considerable gossip and speculation in the early part of the war as he was identified with the Nazi invasion of Europe. Reportedly his house was broken into, rumours circulated that he was going to be appointed as a Gauleter in the event of a Nazi invasion of Britain and that his forests included trees planted in the shape of a Swastika. Some authors have put this down to the fevered imaginings of Dorset locals who could not make the distinction between a cosmopolitan figure such as Gardiner and the foreign aggressors. As others have pointed out, although Gardiner broadcast on the BBC against the Nazis and worked to develop a wartime industry, he remained an ambiguous figure, with actions such as writing to the war criminal Walther Darré in the post-war period. Gardiner exemplifies the ambivalence of some of the pre-war figures that went on to play an important role in the post-war period.

Balfour and the book

Eve Balfour was one of the most active figures during the war; she managed to get her farm at Haughley exempted from the wartime directions to maximize production. It also gave her time to write about her experiment and ideas, which resulted in the book *The Living Soil* (1943), in which Balfour outlined her ideas about how fertilizers damaged the soil, the importance of 'organic' husbandry and her plans to run a farm-scale trial at Haughley. In a Britain, where food rationing was severe and the issue of the domestic production of food was pressing in a manner it had not been for more than a century, Balfour's themes had a new resonance.

Many movements have a founding moment or event; the Soil Association claims to have a founding text. Within five years Eve Balfour's book had reached eight editions, launching, according the Soil Association, the organic movement (Balfour 1943). Balfour brought together much of the work undertaken during the 1930s, with a range of other writings, and introduced a series of organizing metaphors. It offered a way for those who had previously been in the movement to come together in a new alliance. Philip Conford, reviewing its historical importance, described it as: 'a brilliant synthesis of various pieces of evidence from experiments in agriculture, botany, nutrition and preventative medicine, with tentative but potentially far-reaching conclusions about their implications for agricultural and social policy' (Conford 1998, p24). It was not on its own, as others in the movement were putting their experiences into print, during the period of 1940–1945. Through this time of paper rationing, over 20 titles that discussed food, farming and the future from an organic perspective were published (Reed 2004). *The Living Soil* might have become the most well known or influential, but it was part of a broader critical discussion.

Forming the Soil Association

The Peckham Experiment had lost the use of its building – it had been converted into a weapons factory – but it was still attempting to operate. During this period the first contacts between Eve Balfour and the staff of the experiment were made, in the offices of the experiment in London. Ethyln Hazell, who was a medical secretary at the Peckham Experiment and served in the Land Army during the war, recalled seeing Eve Balfour and Sir Albert Howard at Hyde Park Mansions (Reed 2004). It was the staff of the experiment who assisted Balfour in running a 'clearing house' for information about organic farming after the publication of *The Living Soil* (Clunies-Ross 1990). They also conducted most of the administrative work for the founding of her idea for a post-war organic organization – The Soil Association. The importance of the meeting between Balfour and those around the Peckham Experiment was more than just administrative; it provided a new strand to the arguments Balfour advanced moving the discussion towards a farming *and* food movement.

As mentioned above, Balfour's initial response was to consider setting up a 'clearing house' to exchange information about the advantage of organic husbandry. Such a 'clearing house' was set up in Hyde Park Mansions, which then acted as the Headquarters of the Peckham Experiment. The 'Organizing Secretary' of the experiment, C. Donald Wilson, was made available to carry out any necessary clerical work (Payne 1972, p29).[4] Eve Balfour saw the next evolutionary step to be, in Payne's account: 'to start an organisation to run the clearing house, and accordingly wrote off to many of those people who had written to her in response to *The Living Soil*, inviting them to attend a Founders' Meeting, on 12 June, 1945' (Payne 1972, p29). It was these invitations and meetings that were to create the first planetary organization in the organic movement.

In its internal debates the Kinship in Husbandry had noted that although an organization was necessary, no one would have the resources to launch it until after the war. There were a range of groups concerned with farming and food during this period. Although they were different in their aims and structure, there was a sense of a shared project. This is apparent in the personal visits that occurred during the war and the cross-referencing in books. Each was aware of the other. As an example, Bryant in an undated letter (probably 1943/1944) to John Hudfield, director of the National Book League, took him to task over: 'Omissions which suggest – perhaps entirely erroneously – that whoever compiled the bibliography was strongly prejudiced against the organic school of agricultural writer' (Bryant 1943/4). It would appear that a shared sense of identity as indicated by the use of 'we' had started to appear, as people rallied to the 'organic' banner. There would also seem to have been a shared sense of the strategic need for a body to represent this nascent group. Whilst as Friend Sykes[5] recalled, many shared the concern to form an organization to represent these views: 'Meeting my friend Lady Eve Balfour, as I often do, I found that she too had come to this conclusion. Approaching Drs. Scott Williamson and Pearse, we found that they also were thinking along these lines' (Sykes 1946). Mobilization at this time was about finding a vehicle to express their shared concerns and to provide for an internal debate.

Foundation

Its founders are men and women actuated by a sense of service and a thirst after truth. (Balfour 1946, p2)

It was from the 'clearing house' that the invitations to the Founders Meeting were posted. The Founders Meeting of 12 June 1945 attracted around 100 people; it aimed to start a far more elaborate organization than one merely overseeing a clearing house for correspondence:

> *The Founders themselves had one thought in common – the Health of the Soil – from which plant, animal and man draw all their nutriment. They were often men and women of many opinions, some of those strongly held and conflicting. But they knew that the resolution of the conflict lay, not in compromise and war, but in research from scientific knowledge and in education* (Balfour 1946, p2).

The Founders also produced £1600 for the legal fees to set up the new association.[6] It was intended that the new Association would be both a charity and a not-for-profit company. Although a democratic organization, there would be a very British bicameral structure at the top. An elected council would control policy and administration with the members elected in an annual postal ballot, with one-quarter of the council retiring annually, although they could be re-elected. The second chamber would be a 'Panel of Experts' who would supervise the research the Association aimed to undertake. This panel would not be open to election – the council, choosing from nominees put forward by the panel, would fill vacancies. Such a peculiar, oddly British, mix of the appointed and the elected had a lasting effect on the Association. This mix of the elected and expert was only superseded in the 1960s when an advisory committee was established (see Chapter 5). With the adoption of the standards in the 1970s, the experts found a new role in formulating the standards of organic production (see Chapters 7 and 9).

The new association planned to be operational by the autumn of 1945, but the very sophistication of its organization slowed it down. Legal difficulties delayed the inaugural meeting for a year. Payne records this prolonged period in the following manner:

> *In addition [to legal problems], it was necessary to reconcile a wide variety of diversity of viewpoints as to the aims and structure of the Association, and to ensure that the Association was controlled by its members, at the same time as the experts were allowed freedom of control over the research interests* (Payne 1972, p30).

At this stage, the Soil Association lost the Howards' support, a serious blow to their credibility. The Howards had been involved, with Lady Howard on the first committee, but they withdrew shortly before the inaugural meeting on two grounds. First, because the scientific committee was subject to control by the council, which restricted the autonomy of experts, and second, because the Association was too liberal in its policy towards chemical usage. Later the Association would concede both of these points, but only Lady Howard would live to see this. For a short period Albert Howard formed his own society, with many of the same aims as the Soil Association but with his stricter approach; on his death in 1948 this group was combined with the Soil Association. Howard, throughout this period, communicated with Jerome Rodale in the US (see below) and the Compost Society in New Zealand. The detailed substance of Albert Howard's disagreement with the Soil Association remains unclear but he kept up both his global contacts and continued to argue for the importance of his methods.

According to the first issue of *Mother Earth*, the journal that would promote the new association, it would have a threefold mission. First, it aimed to draw together those who were thinking similar thoughts about the natural order; this group, according to Balfour:

> *As an outcome of this interpretation of natural law, they share the belief that the only salvation for mankind lies in substituting co-operation for exploitation in all human activities from soil treatment to industrial and international relations.* (Balfour 1946, p2)

Lest the reader think that the new Association was promoting some form of libertarian thinking, she also noted: 'Disorder and chaos are not natural phenomena. Left to herself, Nature always produces order' (Balfour 1945, p2). Those who were concerned with these matters were to be brought into contact with one another through the Association, as they were 'The nuclei of true constructive effort – these people with vital ideas' (Balfour 1946, p3). The second aim was to foster the development of new scientific knowledge which would demonstrate the complexity and interconnectivity of nature: 'The investigation and interplay of this has its own science called ecology, and it is through the exercise of this science that the necessary research work must be carried out' (Balfour 1946, p3). Third, once in possession of this knowledge the group would be able to inform the public to prevent any 'descent into the abyss'. The public needed to be educated in this new knowledge: 'Not only regarding the need for a healthy soil – the source of our food – but also concerning the way in which that food should or should not be treated once it is grown' (Balfour 1946, p3). The threefold mission of the Soil Association was created. Over the years ahead, the mission would be configured in different ways but the central triptych would remain.

The first council of the Soil Association was appointed, based on those invited by Balfour to the Founders Meeting; it reflected her contacts and those of the Peckham Experiment. Once the new Association was established, it held an election amongst the members to form a new Council and this demonstrated the influence of the agrarian right. The elections brought members of the Kinship in Husbandry onto the council, demonstrating the effectiveness of their tactic to percolate into new organizations. This was not the entry of the far right into the Soil Association, they were already in the first appointed council; rather it was a shift within the politics of these groups.

Lord Teviot and Lord Sempill would have been people whose histories were known at this time. Before his elevation to the Lords, during the 1930s Teviot, as Colonel Charles Kerr, had been warning of a Judeo-Bolshevik plot to overthrow European Christianity. During the early months of the war until his internment, he was a member of the Right Club, a dining club that was supposedly interested in curbing Jewish influence but was actually a front for meetings with the leader of the British Union of Fascists. Colonel William Forbes-Sempill was a landowner, aviator and member of the 'The Link', a pro-Nazi, anti-Semitic organization. Recently Griffiths reports that he has been linked to a spy ring run out of the Japanese embassy, which would have been particularly problematic for him at that time as he was a serving officer in the Royal Naval Air Service (Griffiths 1998, p145).

Perhaps the most disturbing link was the new editor to the Soil Association, Jorian Jenks, who was 'acting as temporary editor to this Association' (Soil Association 1946). Apart from working as a journalist, Jenks was the secretary to the Council for Church and Countryside; an organization that had been 'percolated' by the Kinship in Husbandry, indeed members of the Kinship also helped him to gain the post in the Soil Association: 'Under the sponsorship of Massingham and Easterbrook, I became part-time editorial secretary of the Soil Association in 1945, shortly after its inception, with leave to continue my other activities' (Jenks 1959). Jenks had moved to London from Barnstaple, in North Devon, where he claimed he had quietly spent the war years. In his pre-war career, he had been the agricultural adviser to the British

Table 4.1 *The first elected council of the Soil Association*

Name	Position	Previous affiliation
Lord Teviot	President	
C. Donald Wilson	Secretary	Peckham Experiment
Henry Deck	Treasurer	
Lady Eve Balfour	Organizing Secretary	Haughley Experiment
Maye Bruce	Council Member	
Laurence Easterbrook	Council Member	Kinship in Husbandry
Richard de la Mare	Council Member	Faber Publishing House
Innes Pearse	Council Member	Peckham Experiment
H. J. Massingham	Council Member	Kinship in Husbandry
Lord Portsmouth	Council Member	Kinship in Husbandry
George Scott-Wiliamson	Head of the Panel of Experts	Peckham Experiment
Dr Breen	Chair of Editorial Committee	
Jorian Jenks	Editorial Secretary	Kinship in Husbandry

Union of Fascists (Webb 1976), which had determined how he had spent the previous five years. Until recently, little has been known of the biography and activities of Jenks;[7] however, new evidence means that his role needs to be reconsidered (Friends of Oswald Moseley 2001; Moore-Colyer 2004).

The first elections to the council of the Soil Association appear to have been lively ones; 62 members of the new association stood for election to the new council and 17 were elected. The first edition of *Mother Earth* records that the new Association had more than 5000 members by the time of these elections.

As can be seen in Table 4.1, the result was that four members of the Kinship in Husbandry were elected to the council; Wallop, Gardiner, Massingham and Easterbrook. Teviot moved up from the Vice President to the president and Sempill remained on the council. There was no slate of candidates and it was unlikely that the voting members would have known of the Kinship in Husbandry. It must bring into question what the members believed to be the project of the Association. It also demonstrates another facet to Albert Howard's concern that the council would oversee the scientific panel. What the reaction was of those who were elected to the council at the same time remains unknown, but while the majority of the population elected a Labour government, the Soil Association embraced those who espoused many pre-war ideas.

If the Soil Association was perceived as a right wing organization at the time, by either its contemporaries or its members, it means that much that occurred in the following years can be more more easily explained. Edgell seems to have little doubt that the Soil Association was an organization of the right: 'This was the organization which during its early years had the support of a number of right figures including Rolf Gardiner, Jorian Jenks and Lady Eve Balfour' (Edgell 1992, p629). It would be hard to escape such a conclusion given the preponderance of these figures in the Association. Even if the Association was supposed to attract a range of opinions, an obvious cluster of the far right would have been unacceptable to many people. It must be supposed that some of those within the Association were comfortable with the views these people expressed.

It will remain to be seen in the detailed study of decision-making by the Soil Association's council or from the Balfour archives the exact role that these figures played in the management of the fledgling organization.[8] Certainly the agrarian right held many of the positions of influence and power within the Association and were the largest bloc on the council. The effect on the discourse of the Soil Association was less obvious, as in part they had a pronounced role in forming the constellation of ideas that led to the formation of the Association in the first place. The first evidence of their presence was that many other ideas present in the movement before the formation of the Soil Association were no longer championed. Participatory democracy in the style of the Peckham Experiment disappeared and most of the Association's work was moved to rural areas. The elitism espoused by the Kinship was quickly embraced; few posts in the Association were advertised but seem to have been filled by recommendation during this period. The Association appears to have rejected any ideas of collective or public ownership so in favour at this time. It did keep the planetary focus the empire had brought and the new group welcomed members wherever they were in the world.

Of perhaps greater significance than the people on the council was that they were outside the new ideas and policies that were beginning to reshape agriculture, both in the UK and, increasingly, globally. Debates about agriculture took place in a distinctive manner both within government, and in the foundations and charities that had a central role in developing the new agricultural technologies. The form of these debates came to dominate the strategy of the organic movement for the following 25 years and explains in part why the movement was so marginal.

The green revolution

The organic movement had gathered its ideas and early experiments in a situation where agriculture, particularly in the UK, was neglected, as discussed in the earlier chapters, the focus of the development of agriculture being across the British Empire. This neglect was beginning to be addressed in the late 1930s, a process that accelerated during the war, as Britain became dependent on food supplies from North America. In the pre-war period the major political parties were divided over the issue, with the Conservatives unwilling to embrace change and the Labour and Liberal parties in favour of free markets, as it would ensure, they argued, cheap food for the urban majority. Through a series of reviews during the war, it became clear that there was a new consensus around the importance of science, education and technology as the way to overcome the problems of British agriculture. Perkins, in a history of these events, notes that 'No clear opposition to efforts to increase the technical efficiency and intensity of British agriculture was to be found (Perkins 1997, p197), signalling that although the organic movement was publishing a flurry of books at this time, these were not gaining influence in policy circles and that the movement did not realize what 'intensity' would mean.

The election of the Labour government in 1945 heralded the most important changes to British agricultural and food policies in the whole of the 20th century. Developed by Tom Williams, the Labour Party's most experienced minister in agriculture, it sought to transform British agriculture into a pillar of the economy and cement the nation's role in global affairs. As well as investment in education and research in agriculture, the 1947 Agriculture Act ensured that there was capital investment in farming through the creation of stable markets and assured prices. Farmers would receive guaranteed minimum prices, fixed by the government, to facilitate investment in farming, better living conditions for farm workers, and to raise the supply of food for the urban majority. During the war a series of committees had existed to ensure that farmers followed the most productive practices, and this was continued:

> *The act also contained punitive provisions that allowed the minister of agriculture to direct the technical practices that farmers must use and to dispossess farmers or owners of their lands if the directions were not followed.* (Perkins 1997, p204)

Before the war Williams had favoured nationalization of the land, and arguably through this Act he brought about the nationalization of the technical aspects of farming and the prices for agricultural produce. By encouraging capital investment he also shifted the economics of farming towards the intensive use of machines rather than labour. In so doing he addressed the problem of relationships between agricultural workers and landowners, as the labourers increasingly left the land and the landowners increasingly worked their land on their own.

These policies were partially driven by ideology, as many in the House of Lords thought, and it certainly ended a form of agriculture that had been in existence for 70 years. It was also a pragmatic response to the experiences of the war and the immediate post-war period. The food crisis of 1946 had a deep effect on policy makers at this juncture. As the allied armies had occupied Europe the global availability of food had plummeted; the ration in Germany was down to 1500 calories a day per person, just enough to prevent starvation. The British army was preparing for epidemics that would spread throughout Europe. Simultaneously the situation in India, which was central as a cereal producer and had seen a famine in Bengal in 1943, was more severe than originally thought; food shortages were exacerbating the situation and endangering an ordered withdrawal by the British. The British Cabinet undertook to maximize grain production out of the UK, persuaded the US to take the situation more seriously, including emergency shipments, and reduced the rations of British citizens to ensure supplies for people on the continent. The British managed to persuade the US government of the seriousness of the situation and widespread starvation was averted. For those at the higher levels of government, as well as those eking out their meagre supplies in 1946 and 1947, the precariousness of agriculture must have been obvious.

At this time there was also a growing argument in the US that the country's newly established position would be quickly threatened by population growth in Asia. These arguments were combined with environmental debate about damage already done to the planet's ecosystems. Although these arguments did not necessarily persuade policy makers, they did find an audience in the highly influential Rockefeller Foundation, which was developing new plant varieties. Those working on the Rockefeller programme saw their new package of technologies as addressing an environmental crisis, as described in books such as Fairfield Osborn's *Our Plundered Planet* (Osborn 1948). As Perkin's argues, agricultural policy had come to be dominated by the national security agenda:

> *In the victorious countries, the major analytical framework for building the public policy agenda, with its myriad of issues demanding attention, was the national security framework. It demanded that suggested solutions to all problems, including population and famine issues, would be judged by their estimated contribution to ensuring national security. In the aftermath of the most extensive and destructive war in human history, nothing else mattered.*
> (Perkins 1997, p131)

It was this preoccupation that dominated agricultural policy for at least the next 30 years, and one that the organic movement found difficult to address. A combination of hybrid seeds bred to respond to irrigation and agro-chemicals, often with new farm machinery and technical advice, would be used to create a huge rise in agricultural productivity.

The Rockefeller Foundation had been working to improve yields from wheat in Mexico since 1943, ostensibly with a humanitarian purpose, but by the late 1940s the mission had evolved. Agricultural improvement was not just about holding back or eliminating hunger as an end in itself, but it would undercut the opportunities for communist insurgency in largely rural nations. Agricultural programmes and technology became part of a broader strategy of US foreign policy and the Cold War. Whilst before the war Portsmouth had seen cities as the breeding grounds for communism, in the post-war world and after the experience of China, it was the rural hinterlands that were seen as the heartlands of communism. In 1948 when Harry S. Truman declared his Four Point Program, around which the US would 'fight' the Cold War, point four, 'agricultural productivity', had a central place. High-yielding agriculture became a technocratic component of US foreign policy until the end of the Cold War in 1991. These technologies would be delivered through US aid, and organizations such as the Rockefeller Foundation, throughout developing nations, whilst most in the West followed the strictures of national security to guide their domestic agriculture.

In this context the challenge of the organic movement was superseded, the context and debate around which they had formed completely transformed. The movement lacked the access to policy makers necessary to take part in the debate; its most senior representatives were in the House of Lords and hardly likely to attract the attention of government ministers, especially as the locus of the discussion was now in Washington and not Whitehall. Also what organic agriculture was remained uncertain, as the need for the Haughley Experiment demonstrated. If the advocates of organic farming were so uncertain that they were holding trials they could not match the package of seed varieties, machinery and chemicals that the green revolution was able to deploy. In this context running a research station does have a logic, in that it created an inroad into the infrastructure that was defining the debate about global agriculture at the time. If proof could be generated, if scientists and policy makers could be shown the experiments, then the movement would be able to demonstrate some equivalence to those who dominated the debate.

Advancing via Haughley

The Haughley Experiment would eventually become so closely associated with the Association that it became the main method by which it promoted itself; the means became the end. Such a merger was not unreasonable as the successful trials of the Rockefeller Foundation illustrated; the fusion of practicality and science was beginning to transform the planet. Governments were investing in demonstration farms, field trials and extension agents to spread agricultural practice. Having and controlling a trial farm was to have a voice in the debate, even if it was a small one. Scientific experiments and the goal of gaining the authority that these results would provide became the new strategy of the movement.

This gradual merging of the Association and Haughley took place in several stages, the takeover in 1947 being the first. The second was the financial crisis of 1951, when the money ran out again. It was decided the farms would have to be sold. Only the intervention of Lady Elizabeth Byng, who persuaded her father, the Earl of Stafford, to buy the farms, prevented their loss to the Association (Payne 1972; Balfour 1975). The Earl then leased them back to the Association – rent free. Although this saved their main tactic from ruin, it meant the base of the Association was becoming narrower, rather than wider. To sustain the research station, the movement had become dependent on the patronage of individuals.

Balfour, in her now familiar pattern of creating Rolls-Royce quality organizations on voluntary resources, presented a new possibility for taking the research forward. Conceived in 1951, the 'Ecological Research Association' was formally constituted in 1953 when it took over the Haughley research work. It would, in a style reminiscent of the Rockefeller or Ford foundations, coordinate global research into organic agriculture, with Haughley being only one experiment in a planetary constellation of such projects. It collapsed in 1955 with the Association stepping in to keep the work going. Two years later the Association balloted its members on whether to keep the Haughley Experiment running; the result was overwhelmingly positive. The members undertook to donate £10,000 every year for the next five years to keep it running.[9] By 1960, this fund was falling short and the project was being run from the Soil Association's funds.

The scale of the effort during this period is an indication of the importance that the membership of the Soil Association viewed this enterprise as having, and their financial means. It also reveals that the Soil Association and those supporting it knew that they had to have a planetary remit for their actions, seeking to engage as widely as possible. That they were unsuccessful is hardly surprising; more remarkable is that they even thought they had a chance. The first wave of social movement organizations in the organic arena became dedicated to diffusing not just the movement's discourse but a scientific critique of the dominant form of agriculture and fostering experiments that would refute the new farming technologies.

The Soil and Health Foundation

Across the Atlantic there were a number of experiments in organic farming. The first to sell organic food was Robert Keene, who founded Walnut Acres in 1946, a mail order service. Keene had been to India to teach English and become a follower of Gandhi, and saw organic farming as part of a wider challenge to the values of society, continuing the political project of the Mahatma (Sligh and Cierpaka 2007). At Malabar Farm, the novelist Louis Bromfield was experimenting with sustainable farming during the 1940s (Vogt 2007, p27). The most significant actor was Jerome I. Rodale, who founded the Soil and Health Foundation in 1947. Rodale had become a publisher in the 1930s, with an interest in the promotion of health after founding a successful manufacturing business in the previous decade with his brother. After reading Howard's *Agricultural Testament*, Rodale became an enthusiast and advocate for organic farming and food. Rodale's publishing house became a major source of magazines and books promoting the organic cause, as well as practical methods. In 1947 he founded 'The Soil and Health Foundation' to finance scientific research into the ideas of the organic movement. Rodale brought together an interest in organic farming with a passionate pursuit of health through diet and, to a lesser extent, exercise. A blend of businessman, journalist and campaigner, Rodale was quite different from the British pioneers but it was their influence that started his work for the organic movement.

Rodale read an article by Albert Howard in the late 1930s, which he followed up by writing to Howard and the two began a correspondence that led to the founding of the magazine *Organic Farmer*. Rodale had been publishing a number of magazines in the late 1930s and to focus on his new venture he sold all of the others for $30,000 in 1943.[10] The first edition of *Organic Farmer* was published in 1942, featuring an article by Howard and another by Pfeiffer, with extracts from Charles Darwin's writings. After a few editions Howard was listed as an associate editor. The initial publicity drew only two dozen subscriptions from the farmers of the eastern seaboard of the US. Rodale experimented with editions divided into a section aimed at farmers and another at gardeners, settling on the title *Organic Gardening and Farming* in 1954. By the end of the 1960s it had a circulation approaching one million and was one of the most important vehicles for diffusing elements of the wider goals of the movement (Jackson 1974).

Rodale's work, although not at first widely taken up, did steadily create an audience for his ideas. In 1950 he was called to give evidence to the House Select Committee to Investigate the Use of Chemicals in Food and Cosmetics chaired by Congressman James Delaney, who would later amend the Food and Drug Act to prevent known carcinogens being used in the food chain (Marcus 1997). At the hearing some members tried to bait Rodale because of his lack of scientific qualifications, but he argued that he had a right to be heard as a man of 'general intelligence' who through his work read many scientific reports. Challenged whether he wanted to see the withdrawal of all chemical fertilizers he replied:

> *It cannot be done overnight. I do not advocate it. My purpose is merely to cause the government to investigate and not have these scientists say there is no evidence, and refuse to investigate. They are afraid. There is something hidden that makes them fear investigating this.* (Rodale in *The Organic Front*, Jackson 1974, p107).

This statement can be seen as a call for greater evidence, an impartial call for science to settle the dispute. Yet Rodale is also implying that there is evidence and it is being covered up; in a book written two years previously, *The Organic Front*, he had suggested that the government might force all farmers to use chemicals. He also conceived of the organic arguments in a different manner to the British organizations, bringing food and farming together in the term organiculture:

> *Organiculture is a vigorous and growing movement, one that is destined to alter our conceptions of the farm and to revolutionize our methods of operating them in order to secure for ourselves and others more abundant and perfect food.* (Rodale in *The Organic Front*, Jackson 1974, p121)

That organic farming would result in more abundant food and that it would be more 'perfect', reversed the emphasis of some of the arguments of the British founders whose arguments rested on the damage done to food and the soil by chemicals. In his emphasis on the perfection of food, Rodale also echoes the work that he did in books and magazines to establish vitamins, whole and raw food as part of a good diet.

Rodale, like the British founders, placed his faith in the power of science to make a decisive intervention into the debate. He started 'The Soil and Health Foundation' in 1947 with a view to laboratory work being undertaken on his own farm, something that did not develop quickly enough for his liking. Rodale therefore looked to provide grants to institutions and individuals prepared to undertake such scientific enquiries. The readers of *Organic Gardening and Farming* had in part donated the money for the foundation, so Rodale was able to offer grants for thousands of dollars. The problem the foundation faced was finding institutions willing to take their money, and only a few individuals were prepared to do so, including Dr Ehrenfried Pfeiffer, Professor William Albrecht and Dr Albert Schaatz. Rodale's ideas were so marginal and viewed with such suspicion that even funding that might have refuted his case was rejected.

Vogt argues that Rodale was part of a life and food reform movement, as undoubtedly he was not a practical farmer, even on his own farm supervising the work of others and not being particularly knowledgeable about plants (Jackson 1974; Vogt 2007). Vogt's distinction between an organic farming movement separate from an organic food movement is hard to maintain, in view of the common work and identification of those within the movement. Rodale's work in diffusing the discourse of the movement through his magazines and books introduced and reinforced it to millions of people in North America. The institutions he founded, off the back of this highly successful publishing house, have lasted longer than most of the other institutions initiated during this period.

Pesticides and engagement

> *Some evil spell had settled on the community: mysterious maladies swept the flocks of chickens; the cattle and sheep sickened and died. Everywhere was the shadow of death* (Carson 1991, p12).

That the organic movement fell from public view and influence in the 1950s is not the same as its becoming disengaged or even that it stopped trying, for neither is the case. As can be seen from the discussion above, it was actively engaged in promoting and supporting the research at Haughley, even trying to expand its ambitions. It remained engaged in its ongoing revolt against agricultural technology (see Chapters 6 and 8), and was gathering its forces against the new pesticides. The revolt against insecticides was not automatic within the movement, but it was certainly very close to being so. Although initially suspicious on the grounds of its synthetic nature and its slow deployment, the first stabs against it can be found in 1952. *The Members Information Bulletin* (MIB) introduced Morton S. Briskind's *Statement of the Clinical*

Intoxication from DDT and Other New Insecticides, noting that: 'All members, and particularly any who may, through force of circumstances, consider that the use of DDT is still justified, should procure a copy of this booklet' (Staff-IBM 1952). This opposition quickly moved from the internal discussions of the Soil Association to the public guides to organic agriculture. The iconoclastic Hugh Corley turned on them, as well as rounding on weeds, and organic fanatics:

> *in my opinion a lot of sentimental rubbish has been written and talked about the virtues of weeds. Our task is as always to steer a sane course between unenlightened orthodoxy and 'organic' fanaticism* (Corley 1957, p175).

Even as a scourge of the 'fanatic', Corley will have no truck with poisons:

> *One thing I am sure about. No weed is so serious that the land need be sprayed with poisons to eliminate it – at grave risk to the health and even the very lives of human beings, livestock and wild animals.* (Corley 1957, p180)

Friend Sykes, writing two years later, takes this point and applies it on a planetary scale:

> *What a state of affairs is this! Can we really join the modern scientists in their vainglorious declarations of having drenched our planet in compounds so venomous that the remedies of fifty years ago, even those admittedly poisonous like nicotine, were by comparison life-giving elixirs?* (Sykes 1959, pxxix)

The discourse of the organic movement was beginning to change as it focused on the damage caused by the agriculture it opposed.

Corley's book marks the discursive turn of identifying chemicals as not just poisons for the land but for the people and animals around it. It is clear from his statements that he is not talking about the relatively lengthy process of a poor soil giving rise to poor crops and so ill-health in the consumer. Corley is discussing direct poisoning by the spray or the residue on the crop or soil, whilst the much more institutional or 'on-message' Sykes is still concerned with soil fertility. In his list of concerns, he starts to single out the scientist, rather than the farmer or extension agent as the villain. Although not a full transformation of the discourse, there are indications of future configurations, as the opponent faced and the stake contested by the movement gradually shifted.

In 1959, there was a collision around discussions about organic standards, as a discourse antagonistic to the new synthetic pesticides met with a discussion of retail standards for organic produce. The importance of retail standards is discussed later (see Chapter 7), but it is necessary to consider how it influenced the discussion of pesticides. The attempt to define standards for the retail of organic goods made it necessary to define how such goods should be produced. At that time no common definition existed, as Jenks reflects in an editorial for *Mother Earth* in 1959:

while total abstinence from the use of chemicals is the logical, and indeed an admirable, ideal to be aimed at, it is apt to prove an impossibly high standard for anyone producing for a livelihood, especially at the outset. (Jenks 1959, p10)

The need for retail standards was pushing towards a resolution (see Chapter 7). It entailed an attempt to define what organic farming was, by ruling out what it was not. Increasingly the exclusion of insecticides such as DDT was becoming part of that definitional effort. This attempt at resolution was set against a backdrop of the Association being allied with a major effort to use the law to strike against the widespread use of DDT in the US.

The Soil Association had sent a cheque for $100 to a Miss Marjorie Spock, of Long Island, New York. Miss Spock and her neighbours in the 'Committee Against Mass Poisoning' were resisting the US Department of Agriculture's attempts to use mass aerial spraying of DDT to wipe out the Gypsy moth (Lear 1997). The distinguished ornithologist Robert Cushman Murphy led the group and they fiercely contested their locality being sprayed. One noted commentator said of this campaign of aerial spraying: 'A good many people now have misgivings about the aerial distribution of lethal chemicals over millions of acres, and two mass-spraying campaigns undertaken in the 1950s have done much to increase these doubts' (Carson 1991, p21). The commentator was Rachel Carson, the book *Silent Spring* (first published in 1962). In fostering and supporting this group, the Soil Association was binding itself to a new alliance of groups in revolt against intensive agriculture. They were unable to imagine how that would take hold in the next decade and shake intensive agriculture to its foundations.

Rachel Carson's intervention crystallized a growing movement critical of the indiscriminate use of the persistent synthetic biocides. Carson appeared to embody the critique in her person. An ecologist by training, she had previously written a best-selling work on the ecology of the shoreline. With a background firmly rooted in science, a quiet personal demeanour and no public political allegiances, Rachel Carson became an eloquent spokesperson and symbol of the campaign. Whilst the chemical corporations publicly lambasted her, she knew that she was dying of cancer. After her death, the impact of *Silent Spring* became even more apparent; with its combination of scientific argument, rhetorical construction and eloquence it became a wellspring for the campaign (Waddell 2000). Carson's argument stalled the dominance of the champions of the new chemically based agriculture, and inserted a critique that bound fast to their arguments. After *Silent Spring*, no public discussion of pesticides could be held without her critique being invoked. The organic movement actively supported groups that fed information to Carson, and in turn were buoyed by her work. Rachel Carson, just before her death, contributed a forword to Ruth Harrison's book *Animal Machines* (1964); Harrison was deeply involved in the Soil Association at the time. The organic movement supported the arguments that inspired Carson and in turn were buoyed by her success.

Harassment

Throughout this period many in the organic movement were viewed as 'cranks', 'whackos' and 'weirdos' because of their commitment to alternative diets, and different ideas of health and exercise, as well as their rejection of many of the new technologies. Certainly at a time when tobacco was viewed as at worst benign if not beneficial, the modern classification of nutrients was only just being formulated, and, as discussed above, the struggle to grow more food often equated with patriotism, their ideas were easy to lampoon. Many in the organic movement had railed against vested interests, for some this meant the large corporations that dominated food processing and fo; others it was a secret Jewish conspiracy. The idea that some organized groups were not just opposed to them, but actively working against them became part of the movement's explanations for why events were moving away from them.

Certainly by 1947 articles began to appear in academic journals attacking the science behind the arguments for organic farming (Heckman 2006). Later these appeared in more widely read journals, with R. I. Throckmorton, a soil scientist and Dean of Kansas State College, writing in the *Reader's Digest*:

> *To sum it up, there is nothing to substantiate the claims of the organic-farming cult. Mineral fertilizers, lime and organic matter all are essential in a sound fertility program. Chemical fertilizers stand between us and hunger.* (Throckmorton 1952)

Jackson reports that during this period Monsanto's notes for speakers promoting the new pesticides and fertilizers included rebuttals of organic arguments (Jackson 1974). The new pesticide and fertilizer industries had an interest in, at the least, lampooning the organic movement, such as Rodale's appearance at a Congressional committee, and contesting its scientific credibility. They were often dismissed as 'vitalists', who saw in the soil near magical formulations that could not be proved in science (Geiryn 1999).

Events such as the murder in 1952 of Sir Jack Drummond in France sparked the conspiracy theory that he might have been killed because of his supposed opposition to the new pesticides. Sir Jack had been the Chief Scientific adviser to the Ministry of Food during the war, and had been instrumental in formulating the ration during that period; one of the healthiest diets in history according to its proponents (Fergusson 2007). For some in the movement the opponents of the organic movement had become sinister and dangerous, as the risks associated with their products became apparent. They were seen to be knowingly poisoning people, so what else might they be capable of and what were they doing to advance their goals?

All movements need opponents, defining themselves in opposition to, but also in relation with, their foes. The opponents are often drawn in a sharp binary to all that the movement is thought to be: the local and rooted against the cosmopolitan, the small and virtuous against the large and debauched. At this time the movement's enemies were frequently referred to as 'vested interests' opposed to a movement that was acting selflessly and for the common good. Metaphors such as 'forces of darkness and light', 'David and Goliath struggles' are invoked to express these differences. This is not to doubt the sincerity of those in the movement or their actions, but to show that there are structures to how the opponents are identified, known and portrayed. At this time the chemical companies became a fixture in the arguments of the organic movement, a role they would play for the next 60 years.

Barren years?

In some accounts of the rise of the organic movement, the founders and their works play an important role, and then the late 1960s arrive when the movement gathers force. What those founders did in the intervening 20 or nearly 30 years in some cases is obscured. It has been implied by some that they had all the essential ideas in place by the end of the 1940s, and after that they were only reworked. As this chapter has demonstrated, the years in which some of the lasting organizations of the movement were formed also saw the organic movement fall from prominence. The context in which they were operating changed rapidly and radically, the politics that many of them eschewed were roundly defeated. Britain, once a globally influential player, lost its empire and much of the power associated with it.

Yet the movement adopted a rational approach to trying to spread its influence through an experimental farm. Although much money and energy were put into one small place, had evidence been found, then this might have justified the expenditure. As Rodale demonstrated, he was prepared to fund research that could have disproved his contentions, if anyone would take the money. Running your own scientific experiment ensured that at least the research was being attempted. Attempting to found global organizations that could compete with the foundations spreading the work of the green revolution was hugely ambitious. The goals and ambition of the movement outstripped its ability to organize globally other than through occasional personal contacts, the exchange of letters and the publication of books and magazines.

This period saw not the simple repetition of ideas formulated earlier but a discourse that evolved to account for the new ideas and technologies that were transforming agriculture. These changes became more rapid during the next period – the late 1960s and early 1970s – but it was evident throughout this time. The tensions within the movement between people with different ideas, backgrounds, philosophies and nationality had been held together over a time of conflict and change. It would be the cultural changes of the post-war period that both broke and rebuilt the organic movement. The movement that emerged from this tumult would be one that is more readily recognized in the present movement, but there is a lineage that stretches back unbroken to this time, a time when the movement experimented in how the world might be different without the means or mechanisms to make those changes.

Notes

1   Dorman-Smith served in Chamberlain's government that fell from power in 1940 and Churchill decided that he would not be part of the government. Dorman-Smith spent long periods as governor of Burma in India, until the liberation. He was replaced before the country gained independence.

2   Walter James, the 4th Baron Northbourne (1896–1982) wrote *Look to the Land* in 1940. Frequently credited with coining the term 'organic' farming after viewing the farm as an organism. Steiner initially influenced James but his interests were turned to other mystic beliefs. Like Gardiner he was interested in a range of anti-modern mystical and secret beliefs.

3   Adrian Bell (1901–1980) was a best-selling author of novels and books on country life and farming; he was a friend of Blunden.

4   Payne's work is a particularly important account of this period as she interviewed participants, including Balfour, directly.

5   Friend Sykes (1888–1965) bred racehorses on Salisbury Plain, England; in 1936 he improved his estates with the encouragement of Albert Howard and was a founder member of the Soil Association (see Conford 1988).

6   Approximately £50,000 at current values.

7   Jenks had been interned at the beginning of the war because of his membership of the British Union of Fascists. He may well have been given authority to exercise some leadership of the party by Moseley. Jenks was released after a legal appeal by Moseley led to many interned Fascists being released. The biography of Jenks on Wikipedia exaggerates his national influence in the post-war period.

8   Many of the original files of the foundation of the Association were lost in a fire during the 1950s (Brander 2003)

9   The equivalent of approximately £180,000 at current values.

10 Approximately $370,000 at current values.

5
Small, Beautiful and Reorganized, 1960s and 1970s

Introduction

For most people inside the organic movement and for many commentators the 1960s are seen as the time when the organic movement was born in a wave of countercultural enthusiasm. In part this is because it was a time when many of the current activists and a generation of baby boomers found their own interest in organic farming and food, but it is also because the 1960s has become the common cultural reference point for the start of much of contemporary society. Whilst the previous chapters have demonstrated that the origins of the organic movement lie before this time, the 1960s through into the 1970s saw a transformation of the organic movement as a new wave of people and ideas entered the movement. By the late 1970s the organic movement was very different in form, and in some of its arguments, from how it was disposed in the early 1960s.

The death of Jorian Jenks in 1962 can stand as a starting point for this process of transformation as those who replaced him ushered in a new set of ideas (Staff-ME 1963; Turner 2000; Friends of Oswald Moseley 2001). Through this period the wider environmental movement mobilized, principally around the arguments of Rachel Carson and her book *Silent Spring*. Across the planet a new range of environmental critiques were launched, and as this chapter demonstrates the Soil Association in particular was at the heart of some of these changes, bringing together a new critical community. The practice of organic farming was formalized for the first time, mainly by defining what organic farming was not (Carson 1991; Lear 1997). Organic food also took on a new prominence as novel types of consumption grew, not least off the back of the awareness of pesticides that Rachel Carson had done so much to highlight. Yet throughout this period there were some continuities, many things were forgotten but others were remembered in a new and radical way.

The scale and scope of the transformation of the movement was profound. Some of the early activists in the movement and the critical community died before seeing the new configuration of ideas, organizations and strategies cohere. The arguments for organic food and farming became less about the vital elements within the soil and hence the produce but more about what was absent from organic food – chemical pollutants. Along with this change the strategic arguments for organic food, perhaps paradoxically, became less about science and more about realizing the changes incrementally through consuming organic produce. Throughout this period the key organization, the Soil Association, that had nurtured the critical community so important to the wider organic movement, went through a period of protracted turmoil and at times effectively collapsed, only to be reinvented. This reconfiguration came through the work of three thinkers/activists: Barry Commoner, Edward Goldsmith and Fritz Schumacher. This combination of discursive, generational and institutional change meant that by the end of this period the organic movement was perceived by many to have moved from the political 'right' to the 'left', from the margins to the edge of finding some social and economic power.

The organic movement, I argue, played the role of midwife to much of the wider environmental movement during this period, but in doing so took a considerable toll on its own resources. The Soil Association as the key global organization had both collapsed and been eclipsed. The International Federation of Organic Agriculture Movements (IFOAM) had taken over the mantle of organizing the global organic movement, and did so from a genuinely international perspective, rather than the hangover of imperial responsibility that the Soil Association could at times represent. The near-collapse of the Soil Association was in the end one of its most productive moments. The editorial team at the Soil Association had become the production team for Edward Goldsmith's magazine *The Ecologist*. Through this collaboration new strands of environmentalist thinking were launched, questioning the limits of economic growth and the possibilities for radical social change. Much of the early ideas found in the magazine represented the discussion and thinking of the Soil Association during the late 1960s and early 1970s.

Rise of environmentalism

Rachel Carson's intercession through *Silent Spring* was of course hugely influential but it was part of a resurgence of interest in the environment that had been in abeyance since the end of the war and not its sole cause. Some had written around the same time of the parlous consequences of the new pesticides and other agrochemicals on the natural environment and human health but did not receive the same publicity. Others had raised concerns about other consequences of increasingly intensive agriculture; in Britain Ruth Harrison had written an exposé of the new methods of animal farming and the use of artificial hormones in her book *Animal Machines* (Harrison 1964). Reports of the fallout from atomic bomb testing and an increasing realization of the consequences of increasingly pervasive pollution were gaining prominence.

For many commentators this concern was crystallized and magnified by the colour images of the planet coming back from the space programme. Certainly the images looking back at the planet surrounded by the blackness of space brought with it a sense of the fragility and isolation of the planet. The environmental movement gained ground in groups looking to preserve particular aspects of the environment and this culminated in the first Earth Day in 1969. As discussed in Chapter 3, there had been during the 1930s a sense of planetary environmental peril, but that had been largely concerned with the soil and had viewed the planet in a very different manner. The environmentalism of this period looked to effects that were both unseen but potentially directly fatal and could have profound health consequences for humans. Whilst the environmentalism of the 1930s looked to the degradation of a fragile ecosystem – the soil – environmentalism looked to the pollution of a planet and the humans dependent upon it.

Counterculture

No discussion of this period can fail to take account of the rise of the counterculture, the events of 1968 and a generation that allegedly 'tuned in and dropped out'. The radicalism of this period saw not only a cultural change in mores about dress, relationships, food, music and art but also the rise of revolutionary socialist politics in the Western nations. It may have started in the US in the 1960s but it did not arrive everywhere at the same time or in the same form. The cultural sociologist George McKay argues that there is a continuity from the radicalism of the 1960s through to the protesters of the 1990s and beyond (McKay 1996, 1998). McKay notes that the 1960s were really the 1970s in the UK, in that the events and attitudes that fostered the counterculture in North America only started in the UK in the early 1970s, as free festivals and a 'back to the land impulse' fused with youth culture. Meanwhile in continental Europe this movement took other forms of organization ranging from a deformed version in the shape of terrorist groups in Germany, France and Italy, through to the cultural movements of autonomy from the state in Germany and Denmark or around food in the Slow Food Movement in Italy (Petrini and Padovani 2006).

For the organic movement it saw a new generation become interested in organic food and farming, for the first time since the 1930s. Inevitably this meant that there were some profound differences in dress, values and manners as well as politics. Warren Belasco pictures this as the meeting of older, tennis shoe wearing, sprightly people with long haired, sandal wearing youths – a jarring experience (Belasco 2007). Too much can be made of these differences; the long hair and sandals may have appeared odd, but Rolf Gardiner was still keen on a cloak and leder hosen in the 1960s, whilst Eve Balfour had lived in a commune in her twenties. Each of the different generations had been equally bohemian and 'alternative'; it was their adherence to differing politics and strategies for the organic movement that marked their disparities and this was not always on a generational basis. Rather, the marked shift of the arrival of the counterculture into the organic movement was the wide and popular critique of non-organic food, both of the industrial diet and the manner of its production. Whilst this might have parallels in the inter-war period, the scale and scope of this new analysis was novel; it was not always aligned with organic food but it did bring organic into discussions in a way that had not happened before. It also gave rise to a powerful section of the global food industry that looked to satisfy these desires (Hightower 1976).

The term 'counterculture' is generally used very loosely, so offering a definition is perilous because few may recognize it as what they mean, or meant, by that term. We might consider how one of those who introduced the term, Theodore Rozak, defined it: he argued that contemporary society had become a 'technocracy' that dominated all areas of life with a managerial rationality; 'politics, education, leisure, entertainment, culture as a whole, the unconscious drives and even protest against the technocracy itself: all these become the subjects of purely technical manipulation' (Rozak quoted in Heath and Potter 2006, p32).[1] In such a situation the traditional parties of the left could offer no answer as their resistance had already been incorporated into the system. What, according to Rozak and others, was needed was to strike at the roots of the system, a liberation of the self and the psyche. For Heath and Potter this was a retreat from concerns about material well-being or physical oppression into a search for difference and self-expression which now precisely serves the ends of capitalism.

Another approach to understanding the counterculture is through the resurgence of New Age spirituality that took place during the 1960s and 1970s. In this context the counterculture is concerned with self-actualization, sharing the general argument of New Age spirituality of trying to find a pure or uncontaminated spiritual self within the individual. These counterculturalists are searching for what is authentic and expressive, in order to develop themselves. As Paul Heelas in his study of the New Age movement notes,

> *It is no longer thought necessary to have to 'drop out' in order to 'tune in'. One can liberate oneself from the baleful effects of modernity whilst living in terms of much of what the good life – as conventionally understood – has to offer.* (Heelas 1996, p30)

This idea of the counterculture may even embrace capitalism, in a way that makes the ideas espoused by Rozak appear oppositional, as the self can be realized through money (Thrift 2004).

Often these two different countercultures ran together, along with those who took the notion of questioning the foundations of society as a call to radicalism such as anarchism or deep green thinking or just a dive into hedonism. All shared an emphasis on the self, experimentation and self-actualization that was in sharp contrast to much of the politics and culture that was dominant before this period. Heath and Potter are quite correct in arguing that it departed from the 'old left', although during this period it gathered many who were interested in the 'new left' of cultural, as well as political, opposition. Melucci would argue that this turn to the self is part of attempting to realize the future in the present rather than necessarily a disengagement from transformational politics, as Heath and Potter would suggest.

The Soil Association

In 1962 Jorian Jenks died. In many ways he embodied the link between a particular stream of thought in the organic movement that had moved from being influential, to marginal and finally to forgotten. His successor was a very different person; Robert Waller, a poet, took over the role of Jenks after a career as a BBC producer and latterly the author of a book on Sir George Stapledon (see Chapter 3).[2] The person who appears to have identified Waller as a likely candidate was Rolf Gardiner, who had entered into a lively correspondence with him. Gardiner had made a serious mistake if he thought that Waller was in any way an ideological fellow-traveller. Once in the post, Waller would find himself in a near constant conflict with many in the Soil Association but at the same time he propelled it towards being a more modern organization. Waller recruited a young actor, Mike Allaby, who quickly became engaged in the environmental movement of the 1960s (Waller 1962, 1963).

By the mid-1960s the Haughley Experiment had been running for more than 20 years, and Eve Balfour decided to retire. Seen through the lens of the papers of Rolf Gardiner, this apparently led to a fractious discussion as to the validity of not only the experiment but the goals of the Soil Association itself.[3] A new director, Douglas Campbell, was appointed, who found, according to Mike Allaby:

> *Unfortunately it [The Haughley Experiment] had been managed extraordinarily badly, and round the time, a little time before we had moved there, they brought in a new farm manager a man called Doug Campbell, he had been director of Agriculture in what then was Basutoland, he came in and what he realised very early on was that some of the cattle were close to starvation.*
> (Allaby interview 1999)

Bob Waller claimed that Reg Milton, who had supervised the scientific research, had failed to analyse data accurately and that Ruth Harrison doubted his scientific acumen, concluding that: 'He's been negligent ... but taken the money' (Waller 1966 (approx.)). Haughley entered a rolling crisis; in a letter to Gardiner seemingly from Doug Campbell, the new director stated: 'It has proved impossible for a voluntary organisation to find the large sums required to finance such research annually and it is frivolous to go [on] pretending that it can do so' (Campbell 1966 (approx.)). Barely a week later Balfour had written to Gardiner setting out the differing schools of thought in the council: 'In the present context she [Innes Pearse] believes fundamental research to be the only reason the Soil Association came into being and that without this it has no reason to exist. That is a good deal further than I would go and almost the opposite of your own views I think' (Balfour 1966 (approx.)). The initial alliance that brought the Soil Association into being was being challenged.

The following year brought both another cluster of crises. The farm on which the Haughley Experiment was based, and the Soil Association now headquartered, was sold and then returned, and a 'Scientific Committee' established to oversee the research on the farms. After the earlier intervention of Lady Elizabeth Byng, the farms had been secured for the Association (see the previous chapter). On the death of the noble Lord, the trustees of Lord Stafford's estate had advised that the farm should be sold in the interests of the beneficiaries. A price of £60,000[4] was set on the farm; this was beyond the means of the Association at that time. In the words of an editorial in the *Journal of the Soil Association* (JSA), 'our future was precarious' (Staff-JSA 1967). That was until the intervention of the sandal wearing, Bentley driving, Jack Pye.

Pye had been a member of the Association in the late 1950s, but his membership had lapsed. On a return visit to Haughley in 1967, Douglas Campbell had told him that the farm was for sale. He told Campbell to telephone the trustees immediately and offer £55,000, an offer that was rejected. Pye then bought the farm for the asking price, through one of his holding companies. After years of patronage from the established upper class, the Soil Association's latest patron was a self-made multimillionaire builder. Pye and his wife were greeted in the *Journal* as 'Our new Friends'; this reprieve was viewed as an act of Fate, part of the laws of nature that the Association was trying to unravel: 'The Soil Association has always believed that lasting progress requires patience since it involves the meeting of so many coordinates which are invisible to us before they emerge from their period of root-growth' (Staff-JSA 1967).

Within a short time the Pyes had given the Association an interest free loan to pay off its debts. The infrastructure of the farms was replaced, and Walnut Tree Manor renovated to accommodate the offices of the Association and a library. A cottage on the farm was restored and given to C. Donald Wilson, formerly of the Peckham Experiment. It would appear that one of the Pye's trusts bought an adjacent farm to fatten pigs on and practise organic horticulture (Payne 1972[5]). The Soil Association itself bought a farm it had been campaigning against, on which the previous owners had planned an intensive beef unit (*The Times* 1968). A project to demonstrate the possibility of combining conservation with farming was financed by the Soil Association through a £10,000 grant. Finally the Association had money, and was pleased to be using it. Mike Allaby recalled the money – 'embarrassing amounts of money' – and the pressures that it brought: 'we had to find ways to spend it' (Reed 2004). In welcoming the Pyes, the *Journal* editorial cited the Greek proverb that 'Beggars and strangers come from God'; rather they should have been wary of Greeks and their gifts (Staff-JSA 1967). Payne expressed wonder at where all the money the Association had been given had gone (Payne 1972). With spending such as this, it was not long before even the deep pockets of Jack Pye were exhausted.

The panel of experts who had been charged with giving the experiment and the Association greater scientific credibility had been disbanded in 1962 (see Chapter 4). With Balfour's retirement, a new panel was convened; as Campbell had written to Gardiner, there needed to be a new goal: 'Efforts should now be made to get the SA [Soil Association] entirely out of its publicly-viewed status as a fringe society' (Campbell 1966 (approx.)). Although research remained active at Haughley, the blue-skies experimental approach would be increasingly challenged. There was an important dispute running between those who believed that applied research was necessary to promote organic agriculture and those who held that blue-sky research alone would suffice. Behind the philosophical and personal difference is a question of strategy. Those pursuing the blue-sky approach still contended that a revelation of 'the truth' would propel the organic movement forward to a greater role. For those who wanted applied research, it was that a demonstration effect would inch the organic movement forward, one field, and one stomach, at a time. Gardiner in particular, had been a long-term advocate of the virtues of demonstration. This was a new coalition within the Soil Association: the new entrants, the disillusioned and the long-term doubters.

The final decision to move away from blue-skies to applied research was made. After nearly 30 years the closed system organic section at Haughley was opened; Mary Langman[6] noted the importance of the act: 'Symbolically, the end came with brought-in chicken manure, spread on the organic pasture to increase production' (Langman 1989). Balfour clearly felt this to be a failure, viewing it in no small part being caused by a generational change and part of the general crisis of the Association at that time:

> *In 1970, strange things began to happen. It was as though with a disintegration of a common will and aim in the council a general disintegration set in.* (Balfour quoted in Clunies-Ross 1990, p154)

Mary Langman echoed this later:

> *Many of the younger generation and a few older members as well, felt that it was now far more important for the Soil Association to use the farm to demonstrate productive organic farming.* (Langman 1989, p21)

Framing the debate in this manner had a number of important benefits for those who were to write about it later. The generational argument shifts the discussion from the merits of Haughley to the values of the council, avoiding a discussion of what the experiment achieved. These debates were not only the cold calculation of how to achieve particular ends; the strategy represented the entire careers of some of those taking part.

This sense of general disintegration only got worse when it was realized that Jack Pye had made loans to the Soil Association and not gifts – he expected to see his money back. A storm of recriminations followed, as reasons for this misunderstanding were sought, a solution found and a new order established. Bob Waller went 'walkabout', stricken with grief after the death of his first wife, in 1970 and Rolf Gardiner died at this time after complications from a routine operation. The Soil Association was recreated through the combined efforts of Fritz Schumacher as president, with Balfour returning to play an important role. Rather than reviving the Soil Association as it had been, the organization returned in a form that would foster not just changes in the organic movement but the wider environmental movement.

Emergence

The person responsible for guiding the Soil Association back from the brink was Fritz Schumacher.[7] As Mike Allaby noted, Schumacher was able to be tough when he thought it necessary but he also knew how to frame his arguments for his audience. In a series of interventions, he changed the strategy of the Association, described new purposes and suggested a new set of opponents. To a group of representatives of regional groups he argued: 'Scientific research is no longer our kind of future. We need to develop new lines of co-operation with other organisations in the field; a new pattern of co-operation' (Schumacher quoted in Langman 1989). In a complementary intervention, he repeated his message but attempted to delimit the range of the Association. In the journal *SPAN*, distributed to all members he wrote:

> *Right ideas, in order to become effective, must be brought down and inculcated in this world and the commercial interest must be made into servants of correct ideas. Let us not defend a type of pristine virginity to remain a little esoteric splinter group at a time when the whole world is crying out for precisely the kind of thinking the Soil Association has been engaged in for twenty-five years.* (Schumacher 1971)

Although continuing the radical rhetoric, he emphasized the role of ideals: 'In spite of many experiences to the contrary, the world is ruled by ideas and not by vested interests. Vested interests are troublesome, but they too, are ruled by ideas. Let us work to it that they shall be ruled by the ideas of the Soil Association' (Schumacher 1971). The new path was set out as being in the commercial realm, practical but ruled by moral ideas and seeing opponents as those individuals with differing ideas. The Soil Association was being remade but also increasingly refocused. No longer, for Schumacher, would it be a pressure group active in promoting such a diverse range of alternatives. It would promote organic farming and attempt to sell its produce. The dissemination of the ideas of the organic movement would be conducted through commerce and the practice of farming.

Swiss developments

Organic farming took on a distinctive form in Switzerland, which reflected the influence of the biodynamic movement and also innovations that started there. Steiner started his own experiments at Dornach, which were continued by Lili Kolisko and Ehrenfried Pfeiffer; their laboratories and gardens marked the beginnings of organic farming research in Switzerland. The use of rock powders to restore the balance of the soil and as natural fertilizers had been developed by Julius Hensel in Germany and became popular amongst organic farmers in Switzerland. Urrs Niggli argues that because of the quasi-religious aspects of biodynamics, the federal authorities in Switzerland did not take organic farming seriously until after the war (Niggli 2007).

In the post-war period a member of Parliament, Hans Müller, started to develop a distinctive version of organic farming that diverged from the then-dominant vision of biodynamic farming, echoing Howard in his condemnation of the 'mystic ado' of anthroposophy (Niggli 2007). Müller and his wife Marie began working with Hans Peter Rusch, a German physician and microbiologist, in the early 1950s, attempting to provide a scientific but holistic underpinning to organic farming. Rusch developed a test that he argued allowed the fertility of soils to be determined, a counterpart of the chemical testing offered to non-organic farmers. Müller and Rusch consistently argued against the testing and trials being undertaken by the federal research stations. As early as 1948 these federal stations had started to undertake comparison trials to assess the efficacy of the rock powders used by Swiss organic farmers. Each side appears to have become polarized, but by the early 1970s one of the federal research stations was converted to organic research. Ironically, initially this was largely focused on biodynamic methods as Müller argued that research was not necessary for organic farming, as it was already realized (Niggli 2007). Just as the Haughley Experiment spluttered to an end after nearly 30 years of private initiative, the first state-backed research station, FiBL (Forschungsinstitut für biologischen Landbau – Research Institute of Organic Agriculture) was brought into being. A new wave of experimental farms and research stations would appear in the next decade, but all were acting to support organic farming rather than pursue the scientific mission of Haughley.

IFOAM

Perhaps Eve Balfour was acting on Schumacher's discussion of right ideas as she set off to the meeting that would lead to the formation of the International Federation of Organic Agriculture Movements (IFOAM), although the impetus for the meeting came from the French organization, Nature et Progrès. Denis Bourgeois had been entrusted with the task as a volunteer for Nature et Progrès, an organization which at this time appears to be have been based in suburban Paris and reliant on volunteers working for their rent in a shared house and the chance to grow their own food. The others at the meeting were Jerome Goldstein representing the Rodale Press, Kjell Armani from the Swedish Biodynamic Association, Pauline Raphaely from the Soil Association of South Africa, Mary Langman from the UK and Karin Mundt, who was working for a French newspaper. This group met at Versailles on 5 November 1972.

Roland Chevriot initiated the meeting with the encouragement of Bob Rodale. As discussed in Chapter 4, the idea of a global organization had been mooted in the early 1950s but not realized. Yet again the longevity of Eve Balfour and Mary Langman meant that they were able to carry an idea from a previous period into a time with a very different context. Bernward Geier in his account of the first 25 years of IFOAM sees the spirit of 1968 in the decision not to have a hierarchy but to have a federal structure, with those elected by a member organization and councils responsible for particular actions (Geier 2007). Although less radical, IFOAM was initially registered as a not-for-profit organization under French law. It may well have reflected the spirit of the times but it was also highly practical for an organization with very few resources, already stretched across three continents and with ambitions to link up around the planet, to be so horizontal. Mundt was supposed to do much of the networking in person, but no doubt there already existed many personal contacts that would initiate membership of the new federation (Geier 2007). It took several years before IFOAM gained the resources in both people and finances to begin its role of coordination, but finally a planetary organization for the organic movement was being realized.

A new discourse

Between 1967 and the early 1970s three writers and thinkers were associated with the Soil Association, and in this the organic movement can be said to have materially and intellectually nurtured contemporary environmentalism. This is not to suggest that Fritz Schumacher, Barry Commoner and Teddy Goldsmith were in concert or agreement, rather they described a range of arguments that have become part of the weft of environmental thinking. In such a short space it is not possible to encompass the breadth of their work but rather to suggest their importance for the organic food and farming movement. It is also worth noting their commitment to being public intellectuals and that each of them worked with a team of collaborators who are less well known.

Edward (Teddy) Goldsmith came to the Soil Association looking for practical help in realizing his vision of a magazine that would articulate his particular strand of conservative environmentalism, the ecological limits to environmental growth, the importance of traditional knowledge and technologies as well as a fervent opposition to large-scale corporations. Although Goldsmith's family was very well connected to several influential and wealthy families who had made their fortunes through some of these mechanisms,[8] in launching *The Ecologist* Goldsmith needed money less than he needed the assistance of people who shared some of his worldview and had the technical competence to produce a magazine (Goldsmith interview). As the Soil Association imploded, both Waller and Allaby left to join Goldsmith in his new venture. The editorial and journalistic skills of Waller and Allaby meant that *The Ecologist* was established quickly as a magazine, whilst the financial underpinnings of Goldsmith ensured that it would have a longevity beyond most magazines of the period.

The arrival of *The Ecologist* was announced by the publication of *A Blueprint for Survival*, initially as an edition of the magazine. Penguin, ensuring that it reached an even wider audience, quickly published it as a paperback book. The magazine also became the vehicle for organizing a wider environmental movement. Sandwiched between the references and the appendices was a call for a 'Movement for Survival'; this was envisioned as a coalition of organizations concerned to realize the aims of the *Blueprint* – The Conservation Society, Friends of the Earth, The Henry Doubleday Research Association, The Soil Association and Survival International. The organic movement found itself at the core of the new environmental movement at this moment, a position it would fall from and then later regain. The Friends of the Earth distributed its first newsletter through the magazine, claiming 1,000 members already through the UK, Europe and South Africa; it was aiming to recruit readers to *The Ecologist* to form 'activist cells'.

The *Blueprint* itself was a pessimistic survey of the state of the planet and the pronouncement of the need for a transition to a more benign, environmentally friendly 'stable society' before industrial or mass society collapsed. The driver of this collapse was the tension between a rapidly growing population and the pollution that they created and the finite limits of the planet. It anticipated a short-term future in which ecological collapse and the efforts to avoid it would see widespread suffering:

> *The intensification of agriculture cannot prevent famines within the next fifteen to twenty years, probably affecting parts of Asia, Africa, the Near East and Latin America. Indeed, by causing further disruption to terrestrial and marine ecosystems it must reduce the capacity to support life (The Ecologist 1973, p128).*

Frequently the tone is pessimistic and the challenge presented in the form of charts, graphs and numbers, but the authors urge the creation of a stable society. This stable society will be a community, based on shared values and limited by the environment but in which the arts and sciences would flourish. Against this eco-utopia the stable society was contrasted:

> *A society devoted to achievements of this sort [the Arts] would be an infinitely more agreeable place than our present one, geared as it is to the mass production of shoddy utilitarian consumer goods in ever greater quantity. (The Ecologist 1973, p67).*

To make this change over to a more stable system, inorganic pesticides and fertilizers would be phased out to be replaced by organic versions, as well as the introduction of crop rotations and an end to monocultures in farming. Small farms would take over from large ones:

> *Small farms run by teams with specialized knowledge of ecology, entomology, biology etc, will then be the rule, and indeed individual small-holdings could be extremely productive. (The Ecologist 1973, p51)*

Farming would be a mix between small team run farms and individual holdings that would be able to feed a stabilized population.

Although they went to great lengths to assure readers that the stable society would not be stagnant, the models offered tended to be either from ancient Greece or tribal groups. The tone of uniformity and compliance had at times mechanistic metaphors:

> *For the society to keep moving in this direction, it means that all its members must be imbued with the cultural information that will enable them to fulfill their specific functions as specialized members of their social system. (The Ecologist 1973, p104)*

The absence of crime appears to unconsciously echo the claims of many totalitarian states:

Indeed in such a society, there is no need for external controls of this sort. Systemic controls i.e. those controls applied by the society as a whole through the medium of public opinion, are sufficient to prevent the deviation from the accepted norm. (The Ecologist 1973, p105)

Punishment, restraint and sanctions will not be necessary because 'all citizens will have the good of society at heart'. To survive the coming ecological collapse, the small decentralized communities living within the bounds of the planet would apparently be very conservative and conformist.

Eventually the team that had built *The Ecologist* broke up, with Allaby becoming a science writer and repudiating the idea of 'ecodoom', whilst Waller and Goldsmith continued a looser association. Goldsmith's conservative thinking was broadly at odds with many in the Ecology (later Green) Party and parts of the environmental movement but they could agree on many concerns and his magazine provided a consistent platform for discussion. His views are consistent with a steady flow of thinking in the organic movement in its suspicions of technology and respect for the wisdom of traditional practices.

Fritz Schumacher (1911–1977) led a remarkable life before becoming involved with the organic movement. He studied at Oxford as a Rhodes scholar, and later taught at Columbia University, before returning to Germany for a period at the end of the 1930s. Schumacher had no intention of living under Nazism and returned to Great Britain before the war, but was interned as an enemy citizen at the outbreak of the war. He was sent to work on a farm, replacing the labourers mobilized for the war. Schumacher's talent in economics saw him quickly being used to help the British government's war effort, probably through the influence of John Maynard Keynes who knew his academic work. After the war Schumacher worked as part of the British Army's administration in the reconstruction of West Germany (1945–1950) and later as Chief Economic Adviser to the National Coal Board (1950–1970). During this later period Schumacher travelled widely in the developing world, advising governments and working on practical economic problems.

Schumacher's economic views were far from orthodox in that he was deeply influenced by a spiritual perspective, most famously the idea of 'Buddhist Economics', although by the early 1970s he had been received into the Roman Catholic Church. His spiritual perspective allowed Schumacher to argue that much of capitalism debased humans and the natural world, degrading them in the pursuit of profit.[9] Schumacher's internment as a farm labourer had been an important time for him and he was a keen organic gardener thereafter. For Schumacher all work should serve to dignify those who did it, allow them to express themselves and develop in a common enterprise with others and to engage in production that was socially useful. This work should be organized in units that were large enough to be effective but scaled to allow people to feel a sense of belonging and responsibility rather than being trapped in a concentration of power. This was also based on an argument that knowledge could at best be partial:

The greatest danger invariably arises from the ruthless application, on a vast scale, of partial knowledge such as we are currently witnessing in the application of nuclear energy, of the new chemistry in agriculture, of transportation technology, and countless other things. (Schumacher 1974, p29)

In 1966 he, along with colleagues, formed the intermediate technology group that looked to put many of his ideas to practical effect in developing nations, although in his discussion of Western agriculture, he had clearly delineated ideas.

In his book *Small is Beautiful* (1973) Schumacher considered the proper use of the land, rejecting the distinction between people as producers and consumers, arguing that such a distinction is false and degrading. He provided the example of farmers who produce food that they will not eat because of the pesticides used on them but buy organic food for themselves. When challenged, they argue that they cannot afford to farm organically. This is a separation he condemns because it reveals the wrong philosophical relationship with the land and animals that are part of it:

Man, the highest of his creatures, was given 'dominion', not the right to tyrannise, to ruin and exterminate. It is no use talking about the dignity of man without accepting that noblesse oblige. For any to put himself into a wrongful relationship with animals, and particularly those long domesticated by him, has always, in all traditions, been considered a horrible and infinitely dangerous thing to do (Schumacher 1974, p89).

Once it is realized that people are simultaneously producers and consumers, then traditional economics begins to founder.

As well as condemning the use of pesticides and other new technologies in agriculture, the effects of which he argues are very poorly understood, he also rejects a free market in agriculture goods. It is a question of jointly determining a philosophical position to allow society to be organized outside of the calculus of profit:

There is no need to consult economic experts when the question is one of priorities. We know too much about ecology today to have any excuse for the many abuses that are currently going on in the management of the land, in the management of animals, in food storage, food processing, and in heedless urbanisation. (Schumacher 1974, p96)

Once the values by which we are going to live are agreed, through a correct relationship with the land and the other animals, then humans would recover their dignity and health, whilst the landscape would become beautiful again.

In a later chapter about socialism, Schumacher makes it clear that he is not against socialism, or capitalism. The argument for him is the use of nationalized industries or state controlled industries to deepen democracy, improving people's working conditions, and a more humane administration. And capitalism 'moderated by a bit of planning and re-distributive taxation' (Schumacher 1974, p217) would be able to create an acceptable standard of living. The debate is not whether socialism would be more efficient but whether such measures would develop a better society: 'What is at stake is not economics but culture; not standard of living but the quality of life' (Schumacher 1974, p217). Schumacher looked to new patterns of ownership that would ensure that property was used responsibly not just for the benefit of the owner, and avoided the concentrations that led to organizations too large to serve human needs.

The contribution of Schumacher's thinking and writing to the organic movement, besides his practical contribution (see Chapter 6), was to provide a heterodox mix of ideas that placed an emphasis on the spiritual but also the pragmatic. Schumacher drew simultaneously on Catholic teachings about human dignity, the importance of the poor and communities, with Buddhist teachings about 'right livelihood'. Less obvious, though, was the influence of Gandhi and his associates such as Vinoba Bhave and J. C. Kumarappa, who discussed the importance of an economy of permanence, of small-scale communities and non-violence. The environmental movement has embraced Schumacher but his work is less focused on the environment than it is on the way in which society can be configured to allow people to lead a good life. He questioned not just the green revolution but the also the idea of economic progress.

Barry Commoner was far more of a public figure that either Goldsmith or Schumacher during this period, appearing on the front page of *Time* magazine in 1970. His books *Science and Survival* (1963) and *The Closing Circle* (1971) drew attention to how technology and its products were increasingly undercutting the ecology of the planet. Commoner was a scientist by training, a professor of plant physiology, and chaired the Association for the Advancement of Science. His environmentalism was distinctive in this period as he stood not against technology but for particular forms of it and he came to advocate a form of planned economy to protect the environment. Commoner provided a stream of advocacy for organic farming and the wider environmental movement that emphasized the importance of human health and social justice.

Commoner became one of a group of writers and scientists warning of the dangers of pollution and the environmental crisis in the US in the 1960s. To a great extent they modernized the environmental movement, which focused on pollution particularly after Carson's intervention. Yet, Commoner's vision was more wide ranging and by the end of the 1960s he was a vice-president of the Soil Association. Commoner's contention was that the power of technology had outstripped our knowledge of its effects:

> *The new hazards are neither local nor brief. Air pollution covers vast areas. Fallout is worldwide. Synthetic chemicals may remain in the soil for years. Radioactive pollutants now on the earth's surface will be found there for generations ... Excess carbon dioxide from fuel combustion eventually might cause floods that could cover much of the earth's present land surface for centuries* (Commoner 1966, p28).

Technology had become too powerful to follow a 'trial and error approach' to its use, as with nuclear weapon testing, DDT and nitrogen fertilizers.

For the organic movement, Commoner had a number of important arguments regarding food and food security that made him distinctive in the environmental debates of the time. Commoner re-emphasized the relationship between agriculture and the soil as an ecosystem, as well as the problem of pollution caused by technology. He argued that inorganic (nitrogen) fertilizers were 'an economic success – but only because it is an ecological failure' (Commoner 1971, p150). Only through applying more fertilizer than the plant could absorb would its effectiveness be guaranteed and it was cheap enough to ensure that the farmer would do this, and the excess would run off to impact on others. The same effects he saw in pesticides, which killed beneficial insects and so created an 'agricultural treadmill':

> *Like an addictive drug, fertilizer nitrogen and synthetic pesticides literally create increased demand as they are used; the buyer becomes hooked on the product* (Commoner 1971, p153).

Commoner argued that this was part of the displacement of 'natural organic products' that had taken place since the end of the war, replacing them with 'unnatural synthetic ones' to the detriment of the environment and human health but to the benefit of the economy.

He claimed to admire the business acumen of those who sold these technologies as they had replaced something that previously farmers could get for free through crop rotation and looking after the soil. He was clear that agribusiness, not just the corporations but also the package of business and technology, were major contributors to the environmental crisis:

> *Agribusiness is founded on several technological developments, chiefly farm machinery, genetically controlled plant varieties, feedlots, inorganic fertilizers (especially nitrogen), and synthetic pesticides. But much of the new technology has been an ecological disaster; agribusiness is a main contributor to the environmental crisis* (Commoner 1971, p148).

The crisis might have been different to that faced by the early organic movement in Europe before the war, but Commoner updated and relocated it to the US. He also identified agribusiness as a target for the environmental movement, because of its role in disrupting and damaging ecosystems. Whilst those selling organic food might have been downplaying the impacts of fertilizers, Commoner put them firmly back in the frame.

Commoner found himself in fierce debate with the neo-Malthusians who argued that the cause of the environmental crisis was overpopulation. Paul Ehrlich and Garrett Hardin argued that the rapid rise in population was damaging the environment, and Hardin began to expound the 'lifeboat theory', where developed nations could abandon developing nations to preserve resources. Commoner was clear that to see population as the cause of the environmental crisis, rather than as a factor within it, was to oversimplify a complex problem and attempt to solve it through a biological answer.

> *This, it seems to me, is the main lesson to be learned from both the environmental crisis and the population problem – that if we would survive and preserve both our natural heritage and our own humanity, we must at last discover how to solve, by social means, the social evils that threaten both* (Commoner 1971, p249).

The problem of population was caused by the poverty and oppression of humans, argued Commoner, and the answer was to abolish that oppression and the poverty not the humans suffering it.

For Commoner the answers to the hubris of technology and the environmental crisis were to be found in an active and informed citizenship alongside a form of planned economy. Throughout his scientific career, Commoner had been engaged with public education, and fiercely rejected the argument that scientists had a special role in contemporary societies; rather he argued:

> *The Jeffersonian concept of an educated, informed electorate appears a naive and distant ideal. But at least in one area – science and technology, on which so much of our future depends – an effort is being made to make the ideal a reality* (Commoner 1966, p120).

In the 1970s he was able to argue that this education had made an impact: nuclear testing was banned, pesticides more tightly regulated and DDT banned, and pollution was more tightly controlled (Egan 2007). On the question of the economy, Commoner argued that the economic system of capitalism did not account for the free use of ecological goods that it later destroyed and degraded. He used the example of pollution shortening a worker's life; first the cost was not apparent and the polluting technology appeared to benefit both parties:

> *Later, however, when the environmental bill is paid, it is met by labour more than by capital; the buffer is suddenly removed and the conflict between these two economic sectors is revealed in full force* (Commoner 1971, p272).

The free ecological goods made use of by capitalism included the bodies of the workers and consumers it relied on. After considering and dismissing Soviet style state socialism, he examined the conflict between profit, its relationship with pollution and ecosystem health, concluding that:

We now know that modern technology which is privately owned cannot long survive if it destroys the social good on which it depends – the ecosphere (Commoner 1971, p287).

Production would need to be based on social needs, and in doing so it would be, by necessity, more frugal and egalitarian. Commoner was confident that freed from the need for private accumulation, new technologies would be found that would improve human life and protect the environment.

Commoner's criticism of technology was not the process of seeking to use human knowledge and ingenuity to further human welfare; it was that these were not the uses that the technology was being turned towards. Rather it was the exploitation of nature, with human welfare and health included in that definition, for profit that was the root cause of the environmental problem. An active and informed citizenship in combination with an economy based on meeting social needs rather than the profit motive would see the transition to a more ecologically benign society. Goldsmith's focus on tradition was very different to Commoner's on science and his planning differed from that of Schumacher's decentralized society. Commoner's answer to the environmental problems of the time was that they were caused by a deficiency of democracy and could only be rectified by a resurgence of it. Michael Egan in his biography of Commoner, concludes that:

The apparatus he brought to his activism and the manner in which he sought to define the relationship between environmental issues and a more comprehensive movement for social justice might be worth another – more careful – look (Egan 2007, p198).

Philip Conford, in his book *The Origins of the Organic Movement*, argues that most of the central tenets of the movement were in place by the early 1950s and he places the Christian faith at the centre of the movement's ideas (Conford 2001). This argument would appear to overlook the very important contributions of Commoner, Goldsmith and Schumacher, who did not look back to the writings of the founders of the organic movement but roamed much more widely in their sources and inspirations. They created a re-configuration of the organic movement's arguments that not only suited a changed world but also moved it decisively away from some of the concerns of the earlier group. In the case of Goldsmith and Schumacher this is even more pronounced in that they knew and worked alongside each other as well as the surviving members of this earlier generation and had access to their works. There are continuities in concern and argument, but this is a period of rupture, a change from one period of the movement into another.

Just as the first years of the organic movement saw a collapse of civilization as possible through the undermining of the fertility of the soil, this next iteration saw this calamity happening through pollution and even more damaging practices. They felt they could at least point to systems of agriculture that did work, that there was no need to trial organic methods – they were ready to be widely adopted. The mechanism that would usher in this change had altered; in place of Balfour's faith in science providing a proof, what would be needed was a cultural change. Fortunately for those around *The Ecologist*, the impetus for this change would come from fact that the biological limits of the planet were becoming increasingly evident.

All shared a belief that cultural change was both necessary and obtainable, but what these different streams of thought meant by that varied. Commoner argued for a deepened democracy in which informed citizens could argue and lobby for change. In many ways it was a call towards the old Anglo-American tradition of radical, even revolutionary citizens, stemming from the English Civil War through Thomas Paine and the founders of the US. The call for a planned economy differed from Schumacher's ideas for decentralization, in that it would appear that Schumacher envisaged a smaller role for the state, as he looked to new forms of socialized ownership. Schumacher does not condemn the stifling present in the same terms as *The Ecologist*, rather he focused on the importance of justice and on those who were marginalized in industrial society. Schumacher hoped for a spiritual transformation towards 'crackpot-realism', when people had their own inner house in order and could act in order to promote the greater good. He was quite clear that in order to achieve this, fear would have to be abandoned, which conflicted with the *Blueprint*'s pronouncements of impending doom.

They were united in the importance of cultural change, that through education, contemplation and discussion progress towards a 'sustainable' society would be possible. All thought the need for change was urgent but responded to that urgency differently. Goldsmith continued to publish *The Ecologist* and worked to foster a continued stream not only of interventions into the wider environmental movement but often specifically the organic movement; his intervention in the mobil-ization against GM crops was particularly significant (see Chapter 7). Commoner went onto run for the presidency of the US in 1980, in part an extension of his concern for an active citizenship, and again he later took part in the protests against GM crops. Schumacher died in 1977, and in his honour a number of institutions were formed or recognized their debt to his work: the 'Schumacher Circle' of the Schumacher College in Totnes, Devon, the Schumacher Society in Massachusetts, the New Economics Foundation in London and the Soil Association. This circle became influential in the wider organic movement, with, for example, the Schumacher Society organizing if not the very first, then one of the first, Community Supported Agriculture schemes.[10]

Whilst in social movement terms this trio of thinkers and their associates provided new permutations on the discourse of the movement, and new ways of 'framing' not only the problems the movement faced but also some new goals, they did not immediately provide a new repertoire for how the movement could advance. Schumacher's practical work with the Soil Association is discussed in the next chapter, but despite its practicality, his solution took considerable work before it became a strategy. Their focus on the importance of cultural change did refocus the organic movement away from the importance of scientific proof for organic farming and towards cultivating the social values that would support organic food. In doing so the organic movement provided a cultural critique to complement a range of the political critiques that the wider environmental movement was exploring at this time.

The 1960s and 1970s, in many ways, can be said to have seen the birth of much of the contemporary organic movement, whilst parts of the older organic movement collapsed. The Soil Association was brought to near failure as it wrestled with changing its strategy and discourse. At the same time a whole new generation of younger people looked to the opportunity to farm organically and live in rural areas. IFOAM appeared as an idea, the aspiration of a global organizing structure, however hard that was to realize in any practical way. The energy of this new wave of activist farmers, the ideas they used and brought forward, saw the organic movement make considerable progress in the following decades. They founded farmers' organizations, research stations and a range of other social movement organizations. Much of this activity was overshadowed by the rise of the environmental movement and the organizations that grabbed the headlines and wide-scale support, such as Greenpeace and Friends of the Earth. That the organic movement had provided the network of critique and debate that allowed much of the wider movement to take on new directions became overlooked. It was only towards the end of the century that the environmental movement would rediscover the organic movement and the mutual history they shared.

If this period started with a death, then it also concluded with one, on a TV chat show in June 1971. After entertaining the host Dick Cavett with his ideas on how to live a long life and how his diet had overcome the heart disease he had struggled with as a younger man, Rodale made way for the next guest. During this second interview, host and fellow guest realized all was not well. As medics and staff invaded the set the programme was abandoned; the show was never shown. In less dramatic and fatal ways one group was making way for another, another version of the organic movement was coming to pass (Cavett 2007). That Rodale was considered a good guest for a TV chat show shows that the organic movement had some public visibility and influence; that even in those final moments he was not treated particularly seriously demonstrates the limits of that impact.

The next two chapters consider different facets of that new movement. First, in Chapter 6, is a discussion of the turn that the movement took to selling its produce as a way of providing people with organic food and extending its influence. Initially the imprint of Schumacher is very clear, but over time the influence of the counterculture in its many forms becomes evident. Yet, through intervening in the food market, the organic movement gained more influence and leverage than ever before. The second chapter (Chapter 7) looks to how the political elements of the movement broke out in a mobilization against genetically modified plants and foods – a pathway established by some of the radical thinking of this period but only realized with the assistance of those outside the organic movement.

Notes

1   The social critique of Herbert Marcuse could stand in for that of Rozak, but the latter was more explicit in his call for a counterculture.

2   Bob Waller (1913–2005), a remarkable and generous man poet; soldier, editor and campaigner. Philip Conford's obituary is an excellent testament (Conford 2005).

3   Gardiner's papers are held at Cambridge University and offer a valuable insight into his activities, but obviously a one-sided one.

4 Approximately £900,000 at today's values.
5 Payne's work is of particular importance for this period as she had direct access to the participants only a few years after the events.
6 Mary Langman (1908–2004) was secretary to George Scott Williamson at the Peckham
  Experiment. She was a founder member of the Soil Association, Wholefoods (see next chapter), and an organic farmer (Woodward 2004).
7 Biographical details follow shortly.
8 Goldsmith's younger brother James was a billionaire, involved in many high-finance deals.
  Later he was elected as a member of the European Parliament in France where was associated
  with the political right and environmental causes.
9 Schumacher's spiritual journey had followed the teachings of Thomas Merton, who had created links between Buddhism and Catholic teachings.
10 Based at Robyn Van En's Indian Lane farm.

6
The Rise of Organic Food Retailing, 1980s

In 1972 we could at least unite around the metaphorical 'organic', but now there are many alternatives to that much-bruised name – sustainable, bio-dynamic, GMO-free, and so on – each of which describes a slightly different set of concerns and techniques. And with Wal-Mart a major distributor of organic foods, a conscientious consumer is hard sticking it to the Man by 'going organic' (Belasco 2007, p253).

Introduction

The retailing of organic food began in the 1950s, but it was only through the mo-bilization of the environmental movement in the 1980s that organic food became a regular fixture on the shelves of supermarkets and began to form a notable part of the food industry. The organic movement became part of the global food system during this period through a combination of shrewd public campaiging, private lobbying to gain transnational legal standards, and entrepreneurial drive. This rise to success and prominence also brought new internal tensions about the values of the movement. For some the organic movement was supposed to replace the global food system that it was increasingly becoming part of, whilst for others it was the realiza-tion of their aspiration to combine commerce with campaigning, and yet others were just pleased to have the chance to make money. This chapter will consider this interface between the movement and business and policy networks, as well as the role of entrepreneurialism in establishing a social movement.

Most people first came into contact with organic food through purchasing a product certified as organic in their local supermarket, health food store or perhaps farmers' market. Somewhere on the side of the product will be a small logo denoting the certifying body, and a small amount of text about the agency or regulation that gave that body the authority to do so. Many consumers purchase these products be-cause they believe them to have greater health benefits than their non-organic peers, others because their eating quality is said to be better, others no doubt because it impresses their friends or that it is a fashion they want to be part of. For the retailers, organic is increasingly a 'pillar brand', something that consumers trust and desire, that can give them an edge in often fiercely competitive markets (Cook et al 2007). Additionally, out of sight of the consumers, organic certification provides retailers with a quality assurance scheme that they can rely on, and latterly have emulated in other product lines (Campbell et al 2005).

Whilst most studies of the organic products have focused on them as 'brands', as part of a marketing literature or as quality assurance schemes, within the context of this book they can be seen as a strategy; a strategy that provides a way forward for a movement determined to change the food system that it opposes, develop an infrastructure to support those goals, and to spread their ideas far beyond the networks of farmers, supporters and foodies. The success of this strategy can be seen in the huge growth in those who buy organic products, the public recognition and status of the movement, interactions with the policy making process, as well as the wealth and opportunity that it has brought to many who have taken part in the emerging organic sector. The roots of this strategy start at the very beginning of the movement, but only took on their now familiar contours in the early 1970s, growing massively first during the 1980s and then again surging at the turn of the millennium. Yet this growth, the apparent success of this strategy, has brought strains and tensions that have brought into question the goals of the movement and those working within it.

The first standards

The first self-consciously organic products were produced under biodynamic standards in the 1920s, with, for example, the first organic coffee produced in 1928 (Aschermann et al 2007). A small basket of products does not however create a market, as recognized by the German producer group Arbeitgemeinschaft Landreform in 1938:

> *A first prerequisite for the sale of organic products is to supply the consumer with fresh goods. This, however, in itself has several very important requirements. The quantity must be large enough, it must be available in all seasons, the place of production must make quick transport available to a central point from which distribution can take place quickly and without a hitch. One farm can seldom supply sufficient goods and, above all, supply them evenly over the whole year* (Quoted in Aschermann et al 2007).

Here the problems of scale of production, the need for an adequate range of goods across the year and the importance of cooperation to achieve something approaching a market are laid out. These are the problems that those trying to sell organic products have returned to repeatedly, with the extra complication of having to prove that the goods are organic.

The first organized attempt to retail organic on a large scale appears to have been the appropriately but perhaps unimaginatively named 'Wholefood Store' on Baker

Street, London in the late 1950s. The store was the retail front for the Whole Food Society Limited that had been set up by Lord Kitchener and C. Donald Wilson. Their aim was apparently simple: 'We shall begin by buying the best flavoured and most organically-grown produce we can obtain. We hope to be able to raise our organic standard and later to have food grown under contract in order to raise the standard further' (Wilson and Kitchener 1960). As can seen in the phrase 'most organically-grown' there was some discussion going on as to what constituted organic production; Whole Food provoked a still larger debate about what organic meant in practice.

In 1959 Eve Balfour had returned from a tour of the US and Jorian Jenks reflected an important discussion based on what she found there. Even at this stage Jenks hoped that at some point science would prove the ideas inherent in Howard, Balfour and Pfeiffer's contention that: '"Nature can transmit a quality (of life) which the synthetic substance misses" – the grail quest of Organic food' (Jenks 1959). At that moment science had not found that quality. In the meantime: 'The most important guarantee, it seems to us, that could be given at the present time, is that foods sold as organic, whether 100% or in a lower category, should at no time have been sprayed with any of the toxic, persistent, synthetic chemicals which have recently been so much in the news' (Jenks 1959). Balfour had reported on a system then being used in the US that classified organic produce from a single A through to a triple A, based on how the soil it was grown in was treated and the handling of the product thereafter, with the highest grade AAA ensuring that it was treated in an organic manner after harvest. Although interested in these standards, Jenks was aware that they were not internationally agreed, that a market in one nation could not be sustained, and the system relied on a degree of self-reporting on the part of the producer. The discussion of the early 1950s had not been forgotten but 'She [Eve Balfour] reported that many US members look to the council of the Soil Association for a lead in assembling and collating opinions with a view to reaching an internationally-agreed set of definitions' (Jenks 1959, p10). The Soil Association still saw itself as the global coordinating body for the organic movement.

Wilson and Kitchener at this time were engaged in the struggle to find and sell 'organic' produce. They argued that the full 'straight 100% line' would be too difficult for most practical farmers, and that derivations from it should be allowed. In their examples of the priority considerations regarding purchasing decisions, they placed in order: absence of artificial hormones, antibiotics, open-range, organically treated soil, organically grown food not manufactured or medicated. Acknowledging that this ordering was a change, they justified it by arguing that they considered the use of antibiotics and pesticides 'as being more dangerous than the use of quickly-soluble fertilisers' (Wilson and Kitchener 1960). Nitrogen fertilizers had been displaced as the central element of the discourse, supplanted in the primary position by pesticides. The new hard line would assume organic treatment of soil, but it was to be secondary to the importance of the absence of biocides.

The next innovation was an inspection system, which would be conducted by the Whole Food Society. On behalf of their customers, they would ensure the organic status of the produce:

We believe that we can assure ourselves, before or when buying, that organic producers are not guilty of these practices. But we also believe that most English Soil Association members would prefer to rely on us to make all these enquiries as a regular practice before buying than have to undertake this task for themselves. (Wilson and Kitchener 1960)

The retailer here undertook to assure itself of the provenance of the goods that it bought and to enforce those standards. By 1963 the Whole Food Society was facing difficulties. It had to gain more capital and to reorganize its finances, as a warehouse was necessary and development costs had been greater than expected. Amongst the products offered were oranges from Spain, dried fruit from India, France and California, jams and juices from Switzerland, and vegetables from France (Wilson and Kitchener 1963). A special meeting on the inauspicious Friday 13 November 1963 approved the necessary arrangements. It also restated the aim of the society: 'To attract more suppliers who will provide naturally grown food in ever increasing quantity and variety for an ever widening public' (Wilson and Kitchener 1963, p419). The Whole Food Society was promoting organic produce through a retail led strategy in an urban area. The continuity in the person of C. Donald Wilson links this to the Peckham Experiment as well as the Soil Association. Even whilst the Haughley Experiment continued, what it was seeking to discover was being changed by social definitions produced in the broader organic movement.

This represented a considerable innovation, as what should be excluded from organic products was not only defined but also prioritized, and the basic format of an inspection system constructed. It was burdensome for the retailer, and limited the potential customer to only those stores prepared to undertake these rigours. But rather than having to wait for scientific proof, people could find organic food immediately, assured of its being free from the worst excesses of the new industrial agriculture. Although inspired by efforts in the US it was obvious that the problems that Wholefood faced were those that any scheme with an international reach would confront as they looked to continental Europe and North America for supplies. That meant that any system needed to be straightforward, independent of the retailer and the farmer, whilst being recognized easily by the consumer.

The next development in retailing organic food came in the form of a written agreement by farmers and growers to avoid particular agricultural techniques and technologies, whilst acting in the spirit of the organic ideals. This set of standards was introduced in 1967 in the UK, taking the form of a contract that farmers and growers signed. Although perhaps less cumbersome than placing the emphasis on the retailer it was based on a gentleman's agreement, an honour system. The transition came through an unlikely route via Fritz Schumacher, who, although now remembered as a green thinker and theorist, had made his whole career until that time engaged in the administration of large organizations. According to Schumacher's assistant, George McRobie, during his time at the National Coal Board[1] (NCB) he introduced a quality assurance system for coal (Reed 2004). High quality coal left the NCB depots via distributors and by the time it reached the end consumer, it was wet, burnt poorly and was viewed by customers as low quality. The problem was pinpointed as being the conditions of storage at the distributors. To remedy this,

Schumacher introduced an inspection system backed by a quality symbol.Coal yards that subscribed to the basic standards of keeping the coal well and being subject to regular inspections could display a symbol. Consumers were told what this symbol meant and would choose inspected coal retailers over those who were not. It was this system that Schumacher brought to the organic movement.

As outlined in Chapter 5, during the late 1960s and early 1970s the Soil Association was in crisis whilst the counterculture was transforming the discourse of the movement. Schumacher, through this period, steered the Association towards pragmatism, whilst publicly discussing a raft of new ways of living within the limits of the planet. The 'symbol scheme' brought together the inspection system that Wholefood had discussed with the standards of the trust based scheme and a way of reassuring consumers. To find the new symbol the Soil Association held a competition in which members could send in their suggestions. The final aspect of this system was that, although not cross-border, it did suggest a way that such a trade could be initiated.

As the Soil Association launched its standards scheme, across the Atlantic the Californian Certified Organic Farmers (CCOF) was formed by 54 grower members who founded not only an organization to provide certification of their organic status but one that would also undertake to educate consumers and so expand the market for organic goods, as well as political advocacy. The group had grown out of a certification programme organized by the Rodale Institute in 1972, again demonstrating the role of a core of organizations in seeding new aspects to the movement (Guthman 2004). Although a small organization, the CCOF, like the Soil Association, had seen the importance of setting standards, and a logo, for products that conformed to those principles.

One of the major innovations of this system was that it created the need for a certifying body to organize the inspection system, supervise the standards – and it needed to have the trust of both the consumers and the producers or growers. These new organizations would be dependent on the farmers and growers for their income, but in turn would receive that only if customers recognized the integrity of the standards, and that this was more than just a club. Critics of the inspection system argued that it was often an inspection of the person and their intentions rather than what took place on the farm and that certification was a paper based audit rather than an assessment of actual farming practices.

Whilst certification helped to define what could be sold, the problems of scale, scope and supply identified before the war had yet to be resolved. These are questions that Warren Belasco describes as those of 'infrastructure' in his book *Appetite for Change* (Belasco 2007). He describes a vivid collection of small 'new age' businesses appearing from the late 1960s into the mid-1970s, all catering for the tastes of the counterculture. At this time an assortment of 'health food stores' had appeared to supply those disaffected with mainstream offerings and in search of more wholesome, often organic, fare. Companies such as Rapunzel Naturkost in 1974 – now one of the largest organic wholesalers in Germany (Aschermann et al 2007) – the Ceres Grain whole food store in 1969, quickly followed by Whole Earth Foods in 1974 and Celestial Seasonings in 1970, amongst a wave of others, provided the products for these stores and/or the distribution to them (Whole Earth Foods 2009).

These stores built on an older tradition in Germany, but were novel in the UK and the US, and all provided a vital market for organic goods.

At the same time they also represented a microcosm of the problems that the organic market would face, as some of these stores were determined to remain small enterprises in line with an ethic of environmental responsibility and social conscience. Others were small because they were starting and had their aims set on more than an on going interaction with regular shoppers who, as well as topping up their supplies, were interested in fair trade products and news of local happenings. Belasco cites the example of 'Celestial Seasonings', which pioneered a range of herbal teas that were sold as caffeine free, additive free, and organic. By 1978 it employed 200 people and turned over $9 million but paid its workers less than the local going wage and was increasingly turning its back on health food stores. By 1984 it had been bought by Kraft foods, which was looking for a way into the new and lucrative herbal tea market. Sam Fromartz in his book *Organic Inc* charts a similar process with the soya milk processor 'White Wave' tracing a trajectory from the margins to the heights of corporate capitalism (Fromartz 2006). These are only two examples of a frequent process as brands pioneered in the countercultural milieu of health food stores moved to the mainstream.

Yet these brands were not the same as organic products, which were by definition basic foodstuffs, the appeal of which was that they were not processed or altered in any form. Also there was a conflict with those who developed brands or branded products, as Belasco notes: 'Some New Age entrepreneurs sounded like reaction New Rightists in their denunciation of federal regulation of advertising, processing, and labor practices' (Belasco 2007, p103). Whilst many of either a New Age or New Right persuasion could rail against the role of regulations and urge a new libertarian age, organics could not as it had just constructed itself as the result of a process of regulation. Some would hold that only the organic movement should regulate the standards but this would not protect consumers or the movement from determined fraud. Having some recourse to the backing of the state would add credibility to the creation of organic products as well as surrendering some autonomy.

Legislation

The first legislation offering some safeguards to the organic movement was that enacted by the State of California in 1979. The California Organic Food Act based the standards on those developed by the CCOF, and whilst the state did not undertake to enforce them, it allowed third parties such as the CCOF to seek legal redress against those breaking them. Although only state wide, California's pivotal role in US agriculture and in global food trends, ensured that the Act had greater significance. It also signalled a new departure for the organic movement – that it could and would have to engage with policy makers. After nearly 50 years of acting largely outside the purview of the state, elements of the movement would have to engage with it, but in doing so it would gather some unlikely allies.

The most important legal framework for the development of a global movement has been that enacted by the European Union. As most citizens of the EU know, the process of legislation at the European level often appears to be remote and can also instigate measures that seem to be so bizarre that their original purpose is lost or obscured. It can mean that some member states that are not interested or have not prioritized a topic have to start to enact laws to cover them. This was the broad case for the adoption of EU legislation for organic produce, in that the impetus for the legislation came from Denmark but then every member nation was obliged to start to address the issue of providing a legislative framework. In a stroke, from 1987 onwards, EU countries had to start to prepare for the arrival of a common European wide standard and the emergence for the first time of a common market in organic goods.

The base for these standards was the first international standards agreed by IFOAM in 1980, the realization of a process that had started with the Soil Association in the 1950s. The EU standards started at first with plant material that would come into force in 1991/1992 with the full standard for animal production by 1999. Although these standards would not eliminate debates about particular standards or discussions of the merits of certification altogether, they would change the landscape in which the debates would take place within Europe and all of those who sought to supply European markets. For the movement within Europe it meant that the state had to play a role in the certification of organic produce. The exact mechanism varied between states, although the most common one became a number of certification agencies which could be constituted as charities, not-for-profits or businesses certifying either to the base standards or their own higher ones. Some of these organizations would certify not just organic farming but would view it as one of a suite of compliance inspections they could carry out, for example EcoCert. Others such as Soil Association Certification Ltd, or SA Cert, is a business wholly owned by the Soil Association charity, and inspects to Soil Association standards that are higher than the EU baseline. Suddenly a raft of businesses and bureaucracy mushroomed around the movement, giving it a new institutional weight. This also provided a stream of income that could provide former activists with an income and began a professionalization of campaigning within the movement.

The EU directive paved the way to targeted support across member states, which had been previously left to individual national initiatives, at that point confined to Germany and Denmark. In 1992, as part of the reforms to the Common Agricultural Policy (CAP), it became possible for state support to be given to organic farms as they entered the wider politics of European agriculture. As Alan Greer observes, there is little common about the CAP, so the various nations implemented the support in ways that suited and were adjusted to their agricultural policies or political goals (Greer 2005). But as Heidrun Moschitz and Matthias Stolze comment in their study of the organic policy networks in the EU:

> *although in theory an adolescent bureaucracy such as the EU opens up lobbying options for interest groups, in the case of the CAP the strong involvement of national governments limits the possibilities for civil society organisations to achieve their goals.* (Moschitz and Stolze 2007, p5)

Involvement in the policy networks of agriculture presented a new challenge to the organic movement, as they had to compete or cooperate with the established farming and food lobby.

For those outside the EU it meant that they needed to find a route to access markets. Those in politically organized and well-connected nations, for example Australia, were able to negotiate 'third party import' status so Australian products could enter the EU directly (Lockie et al 2002). Those in countries less able to deal with EU bureaucracy need to be certified by EU-based organizations. The common EU standard did not close the European market to those outside it and arguably began to shape some organic sectors in a way that was favourable to their development (Coombes and Campbell 1998).

The adoption of EU wide legislation jump-started the organic retailing sector in Europe, as member countries in which the organic movement was either very weak or not engaged in the policy process found themselves obliged to set up national regulations for organic production. As many citizens of the EU are constantly aware, there is a gap between a regulation and the implementation of it. Having legal backing, sanctions and a structure in and of itself did not necessarily mean that any farmers would elect to adopt the regulations or that consumers would buy their produce. It did, however, mean that many organic organizations and bodies had to engage with the state in a new way.

Legislation in the US

If the development of legislation in the EU often appeared to be a technocratic process of regulations and committees, the development of national standards in the US was far more confrontational. Through this conflict the US movement gained new organizations and also consolidated a more radical stance. Although, as in the case of California, state legislation had guided organic growing from the late 1970s, the US did not have its own national (federal) standards. This began to change in 1998 as the US Department of Agriculture (USDA) began to outline the legislation that would inform the National Organic Programme (NOP), as had been required in the 1990 Organic Foods Production Act. The initial drafts of the standards were drawn so widely that they would have allowed sewage sludge,[2] irradiation and genetically modified crops to be certified as organic. As Reed Karaim in the Washington Post noted at the time:

> *That's why USDA's organic rules are so troubling. Ignoring the recommendation of a board of farmers, environmentalists and consumers ... In a final twist, the USDA included provisions that could block labels with specific claims such as 'raised without synthetic chemicals' or 'pesticide-free' farm.*
> (Reed 1998)

It looked like the opponents of organic food and farming were going to have the last laugh by wrecking organic standards through legislation. The backlash against the proposed regulations led to 200,000 people registering their objections, a record at that time. Off the back of this protest the Organic Consumers Association (OCA) was formed to coordinate the wider response. The OCA differed from the trade body, the Organic Food Production Association of North America, which had been formed in 1984 and in 1985 became the Organic Trade Association (OTA), in that the OTA seeks to promote organic trade whilst the OCA campaigns for 'health, justice and sustainability'.

Yet as Julie Guthman in her account of organic farming in California makes very clear, for many in the movement that there was any legislation at all was a loss of autonomy:

> For those identified with the organic movement, the federal law represented a huge symbolic loss. It effectively asked agencies that had been most hostile to organic farming to confer it legitimacy, and it forced organic farmers to do business with the very agricultural establishment they set out to oppose (Guthman 2004, p116).

In part this represents the compromises that many other national movements had confronted when faced with legislation but it also reflects the conflictual tone and form of US politics. It proved to be a shrewd estimation of the purposes behind the legislation that eventually came into force in 2002. The NOP was in many ways similar to the European legislation in that it allowed for a plurality of certification bodies, not-for-profits, businesses and charities to be given authority by the central administration. Proof is in the eating, and the NOP has continued to be mired in controversy that has damaged the reputation of organics in North America.

The most obvious example of this reputational damage is a continued series of contamination scandals involving organic food. During June 2006 many people were poisoned across the US by a batch of organic spinach contaminated by *E. coli* 0157 (Wood 2006), or the disputes around animal husbandry on large organic dairy farms (Clarren 2005) and Chinese 'organic' ginger so laden with pesticides that consumers fell ill immediately (Organic-Market.info 2008). This process came to a head in the summer of 2008 when the NOP announced that it had put 15 out of the 30 federally accredited organic certifiers on probation, meaning that they had 12 months to make corrections or lose their authority to certify (Richardson 2008). A less obvious effect was the announcement in 2001 that imported organic products would need certification by a USDA approved organization. This announcement was a powerful way of getting registrations for the new bill, but it did lead initially to a second effect that Lockie and colleagues explain:

> It is fair to say that very few certification organizations from developing countries have either applied for, or been successful in obtaining USDA accreditation. As a result, producers from these countries have been left increasingly marginalized from the US organic market (Lockie et al 2006)

With barely a dozen staff, the NOP's, weak regulation of US organics has had the paradoxical effect of damaging organics through a light touch rather than the heavy hand so many feared in 1998.

Market growth

Generally discussions about the growth of the global market for organic goods have to be hedged with warnings about the quality of the data, as some of it is based on estimates. A swathe of graphs tend to push ever upwards, and the scale of the global figures are hard to gauge against the overall scale of food sales or agricultural production. For example, the estimates for global sales of organic food in 2005 were $30 billion, having risen from $11 billion in 1997. In 2006 31.5 million hectares of land were under organic management compared to 7.5 million in 2000 (Aschermann et al 2007). This is perhaps impressive until it is compared to, for example, WalMart's 2005 online sales of $1.17 billion in a year when WalMart's total turnover was $285 billion. In 2005 the global retail sales of organic produce amounted to just over 10 per cent of the total sales of Wal-Mart. The figures used by many to boost the organic market and the strategy behind it need to be contextualized (Smith and Marsden 2004). They also raise the question of what metrics can be used to measure the success or otherwise of a social movement strategy.

To suggest the scale and scope of this growth it is worth pursuing a case study in some detail to show the dynamics at play within it. The British market for organic vegetables, effectively the first main category of produce about which figures have been collected, was placed at approximately £1 million in 1985. A UK government survey in 1986 placed the market for the entirety of organic produce at approximately £1 million or 1 per cent of all supermarket vegetable sales (Lampkin 1990, p457). A year later the Elm Farm Research Centre put the total value of the sales of all organic produce at £34 million. The Henley Centre trumped this in 1988, estimating the market for organic vegetables alone as being £110 million. The next year the multiple retailer Safeway, then the leader in sales of organic produce, put sales of organic produce at 1 per cent of their total sales. The figures at this time were confused and often confusing, with some suggesting exponential growth, others a more modest tripling. Also sales do not establish whether the prices paid allow the producer to run a viable business, or if the retailer is charging a premium price far beyond what the producer receives. In many ways these figures serve best as an indication of the volume, and in some cases the proportion, of activity.

The confusion was in part the result of the pattern of sales, with multiple retailers selling mostly vegetables, whilst meat and other produce were bought on the farm or through health stores. This made it hard to collect overall figures for the value of sales. Adding to this confusion was that during the peak period of demand at that time, 1989, supplies of fresh vegetables ran so low that 95 per cent of the produce was imported (Lampkin 1990, p456). During this period the number of farms registered as organic grew by 700 per cent in the UK but it was still not able to supply domestic demand. The rapid growth of the UK market led to the development of organic suppliers in its EU partner countries, realizing one of the aims of the common European market, but confounding discussions about the growth of national markets.

The economic downturn of the early 1990s saw the growth of organic sales not only decline but also the capacities underpinning it retract. In the UK the amount of agricultural land certified as organic had reached 1 per cent in this period, but this then fell back as consumers turned away from what they felt to be a pricey product and many farmers let their registration lapse. The growth in this period was in part fuelled by a wave of mobilization of the environmental movement, concerned with the destruction of the rainforests, the impact of acid rain and other chemical pollutants on a diverse range of ecosystems. Awareness of the impact of persistent chemicals in food had been raised repeatedly during this period (Dudley 1991). Ethical consumerism became a vogue for many, with the cosmetic chain 'Body Shop' moving from a small store in Brighton, England to a global brand. The marriage of consumerism and ethical or green concerns did not seem contradictory against the backdrop of the naked greed of the Yuppies and Wall Street. For the organic movement, this mini-boom had demonstrated the potential of the strategy, as growth had been achieved, although it appeared to depend on both a mobilization of environmentalists and an economic boom.

Whether this market would continue to grow at this rate was of obvious concern. By 1990, the market had been growing rapidly for 5 years and it looked set to increase. The Henley Centre forecast sales of £300 million by 1994 and whichever figures these were compared with, this was a massive increase. Forecasting is always a precarious skill, and the Henley Centre was wrong, but only by four years; organic retail sales topped £300 million in 1998. This was caused by the recession of the early 1990s, which saw a slowing of growth if not an actual decline. The first figures the Soil Association provide for retail sales start in 1993 at £100 million, which reflects a fall from the figures of the late 1980s (Soil Association 1998). They also show a fall in the land managed organically or in conversion to organic management, which could be the result of the fall in sales (Soil Association 1998, p7). Europe wide figures for organically managed land reported during this period show that the amount of organic land either fell slightly or at best remained stable (Lampkin et al 1999). The onward march of the organic sector was slowed if not stalled during the mid-1990s.

As the UK rose out of recession into the economic boom of the late 1990s and early 21st century, the growth of the organic market resumed. In the year to March 2007 the retail value of organic foods in the UK was £1 billion, in a market place for groceries of £128 billion, the organic share was therefore approximately 0.78 per cent of total grocery sales (Reed 2009).[3] The emergence of farmers' markets, farm shops and internet-based retailing has seen the scale of these new forms of food retail increase, sales from these outlets rising from £300 million in 2003 to £384 million in 2004 (Soil Association 2006). Taken together, organic food sales were just over 1 per cent of total food retail sales in the UK in that year.

The growth in the British market for organic foods is not one isolated to those islands as it drew in produce from across the world: salads came from Spain, fruit from Italy, dairy produce from Germany, beef from Argentina and Australia, lamb from New Zealand, pineapples from Papua New Guinea, coffee from Central and South America – an organic replication of the dominant food supply chains. As with the first growth of organic produce in the late 1980s, it was in the context of a mobilization of environmental concerns and an economic boom. Yet for all the press coverage, promotion by celebrity chefs and importance lavished on it by supermarket planners, it remained fragile once the economy weakened again.

Converting farmers

This scale of market growth has presented a number of challenges to the movement; not least has been the recruitment of farmers and growers to produce the merchandise demanded. In the first instance the strategy of growing through marketing was to relieve organic farmers of a surplus of produce and to allow them to run sustainable businesses. Yet as the market grew it became increasingly apparent to growers who were not in the movement or who were on its fringes that not only a living, but potentially a very good living, could be made through organic production. Many in the movement were wary of this development, as these farmers might not hold the same values are those already involved in the movement and, indeed, might steer it away from the paths they had followed until this point.

The opportunity to become a farmer or commercial grower differs between countries and reflects many of that particular society's history and aspirations. In large nations such as North America or Canada, land is plentiful and relatively cheap; it is plausible that someone could start farming as if they were starting any other enterprise. However, in many Northern European nations, where land is scarce and relatively expensive, becoming a farmer on a commercial basis is only open to the already wealthy and even a small 'lifestyle plot' requires a considerable investment. Therefore, in some places the movement needed those who were already involved in agriculture to convert their farms to organic production in order meet the levels of demand it had created. This presented a dual problem of how to convince farmers to do this and then, once they had, to make sure that 'they played by the rules'.

As a result, organic farms are not necessarily those founded by people who explicitly did so to farm organically but could be run by people whose families have farmed for generations or decided to do so for pragmatic reasons. Often commentaries, both academic and in the organic media, provided crude categorizations of these farmers, with the 'pioneers' being the most authentically organic and most of the later converts needing to learn to be organic or become convinced of its values. This can be cut across by contrasts between 'family farmers' and 'corporate farmers', the latter always being an embodiment of the system and the former of a Jeffersonian ruggedness that would preserve organic farming from the ravages of industrialism. Wendell Berry, who has both written and worked for the Rodale Press, has taken up many of the themes of these later ideas, including where small, traditional farmers uphold agrarian values of thrift, which are bastardized by industrial agriculture. This perspective is part of a romantic disposition that the first to innovate are the pure and the latter arrivals somehow opportunists – a perspective rarely tested against evidence.

Academic studies of those converting to organic farming have discovered a complex picture. As discussed here, both Belasco and Fromartz could find pioneers who have developed brands and products that have either become major businesses or that have been purchased by multinational corporations. A study that colleagues and I conducted about English organic farmers found an association between new entrants to farming and organic farming, that people entered agriculture to farm organically and that this group were associated with box schemes and farm shops. This does not demonstrate the purity of their intentions in comparison to their peers already in farming but it represents a structure of opportunities (Lobley et al 2008; Reed et al 2008). As one dairy farmer told me, he would love to take his milk to a farmers' market but it was unlikely that he could sell an entire bulk tanker of it. Those farmers already specialized into and equipped for a particular type of production have a trajectory to change. It is also arguably these farmers who have been able to supply the bulk of organic produce. The ability and opportunity to configure a farm business is not an indication of the 'organicness' of the farmers' attitudes, although discussion in the organic media often relies on such simplistic oppositions.

Is it local?

As organic food reached the supermarket shelves it provoked a distinct backlash, as many within and sympathetic to the movement started to promote 'local' food. In many ways this was a return to the early 1970s with its concern for human scale enterprises and food remaining with the social and cultural context in which it was produced. Yet for some it was a direct response to the practices of the multiple retailers and a way of returning control of organic food to growers and consumers. This period also saw a series of innovations in how food could by distributed that made the organic movement a unique area of social and commercial experimentation during the 1990s.

The most significant critical intervention in this debate was the concept of 'food miles', a demand that the total cost of food, including its environmental externalities, was taken into account – requiring a recognition that the distribution of food was a significant part of the energy used in getting food to the plate. It was a direct criticism of the logistics systems operated by supermarkets, that tended to take all produce to a central hub warehouse then distribute it out to the various stores in its chain. In combination with increasingly sophisticated just-in-time inventory management, this meant that a significant amount of stock was actually warehoused in trucks on the road at any one time. These trucks were often returning from their runs completely empty, so that the number of miles any item would travel could become exaggerated as it was shuttled from its site of production, to the central hub and then on again to where it would be consumed. The argument was simple; why not cut out the central hub and take it direct to the store?

Other factors influenced the turn towards local markets and systems of distribution. The immaturity of the organic market meant that wholesale markets were often not well developed and the number of routes to market limited, particularly for small producers. Some producers who had a route to the supermarkets found that their attitudes and demands were antithetical to organic ideals. This ranged from arbitrary aesthetic guidelines that ruled out large parts of the crop, for example 13cm as the maximum length for courgettes or the abusive attitudes of the buyers. Practicality merged with values as organic farmers and growers looked to ways of taking control of their own distribution.

North America provided a number of distinctive examples that were directly emulated by others in the organic movement, although in that translation aspects of it were adjusted. Farmers' markets offered a novel twist on the older tradition of street markets and the goal of farmer-to-consumer interaction. It also suggested a certain scale and product range by the farmer for them to find a relatively local market as a worthwhile outlet. The community supported agriculture (CSA) or subscription model was another. In some instances people paid a lump sum to the farmer or grower representing a share of the total crop, in others they paid a weekly amount. In this model the grower or farmer is working with a defined group of people, bypassing any form of market interaction. Selling through a farmers' market, farmers cannot guarantee their sales and have to take time off the farm to sell their produce; a CSA scheme allows them to focus on producing the crops and provides a stability of income. The disadvantage of CSA was that it required large enough group of people to be recruited to make the shares both something that a family could afford to purchase and be expected to eat.

As these innovations began to diffuse through the organic movement, a new synthesis appeared: the 'veg-box scheme', which was pioneered in Exeter, England. Tim and Jan Deane established a group of customers who agreed to pay a fixed price each week in return for a box of vegetables produced on their farm. Customers received a box of vegetables that reflected what was available on the farm rather than what they might have chosen, but at a lower price than they would have paid at a greengrocers or supermarket. The grower in turn had a simple distribution system that helped with their financial, and planting, planning. Although the Deane's scheme reached only a few hundred families, other schemes start to reach tens of thousands as farmers and growers emulated and modified the scheme.

The most high profile box scheme became one based near Totnes in England, close to the Deanes. Taking the model of the box scheme from the Deanes, Guy Watson rethought the system to provide for a new scale of distribution. Based on the produce of a cooperative of ten farmers in the area, plus items that could be bought wholesale, Watson offered franchises for particular areas in the south west region of England. In this way growers and farmers focused on production, Riverford on marketing and running a central distribution system, whilst the franchisees took on getting the actual boxes to the customers' doors. Another British scheme also provided a large-scale version of the model. Abel and Cole acted as a wholesaler, packing and distributing boxes in London and the south east of England (Heeks 2007). The early years of the 21st century saw a mushrooming of box schemes, as customers found them an effective way of bypassing the supermarkets, or just convenient (Purdue et al 1997).

Although none of these methods of distributing food, selling shares in the crop, delivering to the customer or running a market is historically novel, in their contemporary configuration they are, in the sense that each of these systems speaks directly to the discourse of the movement – through a closer link between the consumer and producer, the ecological provenance of the products and embedding of new social relationships around food. Each method has had to struggle with the practicalities of realizing an organic food distribution system, particularly across the axes of scale and localness. Yet each is closer to the aspirations of the movement than the multiple retailers. Farmers' markets have found a direct way of connecting the farmer with the consumer, although organic may not be as prominent a connection as once intended (Kirwan 2004; Moore 2006). CSA schemes have brought new and relatively tightly bounded networks together, whilst the model of Riverford has found a way of bringing together a cluster of small businesses, in a way that allows them to cover a larger territory than they would otherwise.

Michael Winter, in his discussion of the rise of local foods, has pointed to their role in creating a 'defensive localism', as rural communities have rallied to protect their 'own' farmers (Winter 2003). In many ways these schemes allow people to defend their 'local' organic grower, sidestepping the multiples altogether. For the organic movement, the debate around local is more serious than parochialism alone; it highlights tensions within the discourse. The environmentalist maxim of 'acting locally and thinking globally' is potentially destructive in a movement with planetary ambitions. In its crudest form, the demand for local food naively 'trumps' the process or need for organic certification (DuPuis and Goodman 2005). In its ideal form of the early 1970s this was supposed to be about intimate, intertwined communities, but it can often serve as a way of not discussing the actual practices of farming and foregrounding only the importance of local provenance.

In its more complex form it begins to undermine the networks of trade and association that bring communities together through the exchange of food. This can be articulated through calculations of carbon footprints or lifecycle analysis where the environmental costs of raising crops in naturally temperate climates are balanced against growing the crops in colder climes. Equally, it can be expressed in providing the opportunity for poor communities to access markets for premium products. This access can be cast as levelling the economic playing field or solidarity towards those who through the lottery of birth have not had all the advantages of life in a developed nation. Local sits on the fault lines in the organic discourse between the arguments for a wider societal transformation and those that look to prevention of both environmental damage and social change. The contest between social change and social conservatism is a constant in the organic movement. This tension was apparent in the early years of the movement and continues to be so, not only in the retailing of products but the policies and politics of the organic.

Multiple retailers – a Faustian pact?

Part of the new radicalism of the organic movement forged in the late 1960s and early 1970s saw organic food being opposed to the multiple retailers or the 'not-so supermarkets'. Organic represented small and sustainable, whilst supermarkets represented the big and ecologically damaging, the human scale versus the machine. Yet it was supermarkets and their supply chains, whether rooted in the counterculture such as Whole Foods or Trader Joe's in the US, Macro Wholefoods in Australia or in the global multiples such as Carrefour or Tesco, that brought organics to a mass market for the first time.

The first experiments with organics in supermarkets in the mid-1980s met with some difficulties as they did not conform to the aesthetic standards demanded by these retailers. This conflict between the movement's demands for quality being an internal factor and the multiples' focus on surface appearance above all else has never been resolved, although each side found ways to get the produce on the shelf. In the US a chain of supermarkets developed around an organic offer, whilst in Western Europe they found their way onto the shelves of existing chains. During the 1980s they were obvious experiments but, with the advent of EU standards in the 1990s, the multiples found the inspection systems conformed to their ideas of quality. Multiple retailers are highly reliant on quality assurance schemes to protect them from substandard or dangerous produce, monitoring the process of produce from farm to plate (Busch 2000). Most consumers are unaware of these schemes, but organics offered multiples one that their customers were not only aware of but for which they were prepared to pay a premium.

For retailers organics offers, as well a chance to create profit, the opportunity to play complex games with the range of products they offer to customers. One market research company I interviewed spoke of a major British supermarket using organics to create a 'green halo' through which the organics stocked in the store influenced customers' perception of the other goods in the store (Cook et al 2007). In the intense competition between the multiples, organics have been used to position chains in comparison to others. The British chain Waitrose trades on the quality and ethics of its offer, which is heavy with organic produce. In taking an aggressive position in selling organic in 2006, WalMart argued that it was going to increase the affordability of organics (Lockie 2009). Significantly, Tesco describe organics as a 'pillar brand' and as such an important part of the roster of goods their stores sell. This highlights, as in the discussion above, a tension between organics as a brand and a production system.

As the credit crunch turned into a global recession, the sales of organic goods stalled. Organic sales have hit problems before – the economic recession of the early 1990s saw the sales of organic food and the number of farms fall in many markets. The trajectory of this fall is significant; when the recession began, the media started to run stories on the decline of the sales of organic food, rebutted by retailers and organic bodies, but an obvious story. These stories positioned organic as a luxury, an unnecessary expense that people could and would ditch as soon as they had cause. To be fair to the journalists, many organic products had been positioned in the market as luxurious, top quality products at top quality prices and so the charge was well placed. In the early months of 2009 it became clear that sales were falling away, as consumers generally 'traded down' from luxury brands to more basic ones, from joints of meat to mince. Organic farmers found that years of investment and work was suddenly vulnerable, if not lost entirely. After years of organic figures emphasizing the aggressive growth of the sector, of graphs that pointed ever upwards, decline became a very real possibility. These events revealed the weakness of the strategy; if organic is of such great importance to the sustainability of the planet then surely it is too important to be left to the vagaries of individual purses and wallets?

Past its shelf life?

In the debates of the Soil Association in the early 1980s, as it reorientated its strategy towards winning over the market, Eve Balfour rallied support for the original mission of the Association. As Eric Clarke explained:

> *I do not believe, for example, that the Soil Association should take as a major activity a policy to persuade consumers to buy the products produced by some of our members. There may well be a useful place for such a campaign, but it should be carried out those organisations specifically orientated those interests* (Clarke 1982–1983).

By giving up the research activities of the Association it would become only about promoting the interest of some in the movement, advancing the means of advancement rather than its ends. In one of the first academic papers on the economics of organic farming, David Bateman and Nic Lampkin pointed to the goals of the strategy, which was not the whole of British agriculture but:

> *The question that is being posed here is much simpler: would it contribute to the achievement of the objectives indicated earlier if a significant part of our farming became organic* (Bateman and Lampkin 1986, p93),

The goal of the strategy was clear, that a 'significant part' of agriculture would become organic through consumers and farmers working in tandem. Whilst the extent of 'significant' remains a proportion of debate, it would seem reasonable to think that it is greater than that which has currently been achieved.

By 2008 the strategy of growing the reach of the organic movement through the labelling of products, supported by a system of inspection and certification had developed the strength and reach of the organic movement. If, as Warren Belasco argues, the role of the organic movement has been to provide the counterculture with the infrastructure necessary to change the food industry, it now constitutes a considerable industrial sector in itself. What Belasco describes as the 'infrastructure' is far larger and wider in its scope than it has ever been in the past; some people have made small personal fortunes, others managed to buy their farm and others to live a life they have dreamed of, all thanks to this strategy. The organic movement had travelled a very long way from the early speculations by Albert Howard of the possibility of selling organic produce separately. Yet according to its critics, even those within the movement saw it being undercut by the demands of the strategy. For those critics the standards of production were being lowered so that, in fact, the claims for organic farming being more sustainable are less substantial than they should be and that many people who are not interested in the aims of the movement have been sucked into the organic sector and have from there been undermining those goals. In being seen as just another 'brand', the wider ecological and social messages of the organic movement have been lost. By becoming a niche and quality brand, often the most expensive, it has allowed organic to become associated with privilege rather than an answer for the majority of the population.

Knowing the market/movement

Relations within the organic movement have been transformed by this strategy as a raft of professional organizations have found a role within the movement, along with a transformation of how those organizations relate to others in the movement. Several models of how environmental groups relate to their mass membership are current; one example is that adopted by Greenpeace from the 1970s onwards. This model has a large number of supporters, most of whom are engaged through paying a membership subscription and are 'serviced' by mailings and magazines (Eyerman and Jamison 1989), whilst a small group of highly trained and professional activists undertake either high profile protest activities or technical lobbying work on behalf of the largely passive membership. It is tempting to see this as an analogy for the activities of the organic movement, as discussed in the next chapter.

Increasingly the only substantive link between the professionals in the SMOs, particularly those dominated by the business of certification or promoting organic food and farming, is through market research. Arguably this is only part of a broader trend in politics where marketing has supplanted campaigning, but it shares the same fundamental flaws (Moloney 2006), that those in the movement – whether they be farmers and growers needing certification, members of lobbying groups or consumers – are related and imagined through the lens of marketing. The daily or weekly act of consumption is not the same as being a citizen, of partaking in the civic life of a society (Trentman 2007). Market research asks questions based on assumptions about human behaviour, often the very norms that social movements seek to oppose. In this way organic can be discussed as a 'brand' by those who work within the movement, as the language of the market colonizes the very organizations that sought to transform it. Rather than knowing the movement through being part of it, the professionalized social movement organizations act upon the movement through the tools of marketing rather than in dialogue and discussion with it.

Recent discussions about the marketing of brands and the limitations of ethical consumerism have found new areas of convergence. It has become increasingly apparent for many companies controlling brands that dedicated consumers both develop and will contest changes in products (Arvidsson 2005). Hayagreeva Rao in his book demonstrates how social movements and activists can contest the introduction of new technologies such as GM plants or boost them as they did for cars (Rao 2009). So control over the brand is not as absolute as many marketing people maintain; it involves negotiation and interaction that is beyond the activities branding consultants anticipate. In their trenchant critique of the relationship between the ideas of the counterculture *Rebel Sell* and contemporary capitalism, and Potter Heath argue that the ideas of being an individual, of cultivating difference, play into the hands of contemporary marketing rather than opposing it (Heath and Potter 2006) – this is an argument I will return to in the final chapter. If major corporations cannot control the brands that they have created, logically there is no reason to suppose that the organic movement can control the distributed and contested 'brand' organic.

The organic movement has in part been shaped by the way in which it has interacted with the different institutions it has encountered, in both the state and commercial realms. In Europe the process of becoming regulated was in many ways brought within the orbit of a broadly social democratic state. This process of becoming entangled with the institutions of the state has generally been consensual, with obvious exceptions such as the dispute over GM crops (see the next chapter). For the most part, the organic movement remains in control of the regulations, even if in discussion and occasional tension with the organic industry. The very strength of these regulations has become a useful tool for the broader movement in debating the direction that they should take (see Chapter 8). Hilary Tovey, in her study of the Irish organic movement argued that it had become institutionalized by being drawn into the existing networks of agricultural policy (Tovey 1997).

Julie Guthman, in her conclusions about the development of organic agriculture in California and how it has become dominated by the demands of the organic industry, looks to how things could be different. She suggests more technical support, stronger regulation of pesticides in particular, and carefully targeted subsidies – particularly for those farms in transition to organic farming. Guthman argues that the neo liberal paradigm of the state needs to be rolled back for organic or alternative agriculture to flourish: 'because of it redistributive capabilities, only the state has the capacity to unlock some of the mechanisms of agricultural intensification' (Guthman 2004, p179). In Europe the state does frequently intervene in agriculture to prevent many of the problems Guthman identifies and the organic movement has been a beneficiary of that tendency. Yet, as Tovey argues, there is a price to be paid for being intertwined in the institutions of the state, as the experience of US federal standards demonstrates.

The criticisms of Guthman and Tovey need to be taken more seriously than many that have been launched by those who are romantically inclined to view the answer to be small-scale gardening or some return to a pre-industrial state (see Chapter 8). Through coupling campaigning with marketing, the planetary organic movement has achieved far more in the last 30 years than it had in the preceding 30. Ironically, because of the wealth generated by the organic sector and its inclusion within agricultural policy networks, more scientific evidence about the advantages of organic farming and food has been produced than the isolated work of Balfour or Rodale could ever have achieved. This evidence has underpinned the policy support that organic farming has received, but has not provided – and is not likely to provide – the overwhelming evidence the early movement hoped for. As will be apparent in the next two chapters, organic farming and food has become part of the debates about the future of farming and food in many domestic and international arenas. It has yet to make the impact, however, that those who originally constructed the strategy hoped for.

This tension between the market and movement is perhaps best captured in the figure of the entrepreneur. In the discourse of the movement the ideal farmer is frequently Jeffersonian, a stalwart of civic society farming with a responsible attitude towards the land and her fellow citizens. Yet this farmer is also an entrepreneur, selling her produce in the market and seeking to make a justified profit. Social movement theorists write of movement entrepreneurs, the activists who work to build and sustain a movement, taking opportunities and innovating. Then there is the entrepreneur who creates a product, frequently now a brand, and uses that to either create or enter a market to realize a substantial profit. These profits may be used to gild their nest or in pursuit of the greater aims of the movement, or both. Quite which is the one that the movement wants to see at any juncture is contested even within the movement.

Frequently the elite of many organizations of the organic movement are made up of people who fall into one of these categories, the Jeffersonian farmer, the movement entrepreneur and the brand entrepreneur. The wider movement embraces all of these possibilities but this does not mean that there are no tensions in this alliance. Having kept those tensions in check, configured so that they generally work together, has allowed the movement to advance. Yet, as this chapter has demonstrated, that configuration is contingent and requires careful maintenance. Rapid shifts in power or changes in the context in which the debates about food are taking place could see this alliance unravel. The next chapter demonstrates how the interests of these different tendencies came together in a campaign, whilst the final chapter points to tensions that could see that alliance unravel.

Notes

1   The nationalized company responsible for coal production in the UK from 1947 to 1987.
2   The use of human sewage has a long and contentious history in the organic movement. Many pioneers pointed to how it closed the circle of fertility. Municipal human sewage is likely to be contaminated with a wide range of cosmetic and cleaning products, some of which contain toxic chemicals. As the debate continues, it has come to revolve around the treatment of sewage before it is used on any agricultural land.
3   By this time organic cosmetics, clothing fibres and herbal products form part of the market, so comparisons of food alone may be misleading.

7
Fighting the future – against GM crops

In the summer of 1998, just over ten miles from my home I watched a local organic farmer dressed as the grim reaper start to scythe down a trial plot of genetically modified maize. As the uniformed police officers chased him around the field, even more of the crop was destroyed. Alongside me at my vantage point the special branch[1] captured the whole escapade on video, which no doubt the police helicopter did as well. More importantly for the farmer, as he was being arrested and escorted to the waiting police van, the press photographer captured it as well. Within days his picture wearing a skull mask, flowing white robes and carrying a scythe was appearing at first across newspapers and then their websites, then activist sites and later it found its way onto posters. In a few, apparently simple, actions the organic movement had taken a different trajectory in the form and diffusion of its protests. From a field in rural Dorset, England, the farmer engaged in a global dispute about the future of farming, not just through his deliberate physical act of protest but through the imagery of the protest, a clear example of the links between physical actions and discourse (Reed 2001).

The mobilization against the introduction of genetically modified (GM, or genetically engineered, GE) plants saw the organic movement rise in its first united public protests and take its place within the broader environmental movement. During this process it was challenged and changed, with new groups entering the movement, radical paths being charted and its role in the politics of food firmly established as never before. The mobilization demonstrated where the power of the global organic movement lies and where future battle lines will be drawn. Through the protests, the organic movement became involved with a diverse range of allies and with new forms of public activity that had not previously been in its repertoire. The full effect of the processes initiated by these alliances, and the protests they unleashed, have yet to be digested by the movement.

In this chapter we will explore the context in which these protests started, and the way in which the organic movement became the centre of both the arguments and in many instances the actions of these planetary protests. As with any globalized protest, it is difficult to capture the full breadth of the actions taken but we will track the protests from Germany, to the UK, briefly across to France by way of a new global movement of peasant organizations and on to the US, before returning to the UK. Ultimately both sides in this conflict have been able to claim victory, in that GM crops are grown widely across the world and play a significant role in most commodity crops. Yet in one key global market, the European Union, they are widely shunned by consumers and farmers alike, demonstrating the limits of the power of the protest movement in that they have not been able to stop the technology but they have brought it firmly into question. If the introduction of GM technologies was the next stage of the green revolution, it was at best partly successful and its final ascendency is not yet assured.

In the discussion of how people become involved in protest activities or mobilized, there is an ongoing debate about the relative importance of existing social networks and the role of the media in providing people with the motivation to take part, described by some as 'cognitive frames' but broadly, in this book, as discourse. Proponents of both approaches have carefully researched evidence to prove their perspective; without anything approaching that amount of detail, I argue in this chapter that both were evident in the mobilization against GM crops, because of the context in which the protests took place. Those already in the networks of the organic movement and indeed the wider environmental movement were aware of the threat to organic food, whilst those outside these networks were given a 'moral shock' and a simple way of participating. The breadth of the protests offered a range of possible actions that were diffused across borders as the intensity of the protests flowed and then ebbed (see Chapter 2). As the protests progressed, the range of targets that were contested alongside, and through, these plant varieties spread, offering the organic movement new avenues of influence and activity, as new allies joined the mobilization.

Context

The last days of the second millennium were certainly anxious ones, as the days ticked towards the predicted global computer hitch of the millennium bug and expectations grew of an overdue global flu pandemic. In Europe at least, these epochal signs were placed in a disturbing context by the confirmation in 1997 of a link between mad cow disease (more correctly, Bovine Spongiform Encephalopathy (BSE)) and a variant that had appeared in humans. Predominantly associated with the cattle industry in the UK, it had first been identified in the early 1980s but for nearly a decade any possibility of its crossing over into humans had been denied. The announcement in 1997 that a link had been established to BSE, and that a few people were already ill with an incurable degenerative brain disease caused widespread consternation (Dealler 1996; Ford 1996). Serious debate saw the potential death toll rising towards half a million in the UK alone. With the once global trade in British beef in tatters, the UK government set about eradicating the disease in cattle, with their toxic remains eventually being burnt in power stations to ensure its complete destruction. Britons and those who had been resident in the British Isles during this period found that they were unable to donate blood outside the UK and were treated in many hospitals as a potential source of infection. The once global British beef industry shrank to operating only within its home islands. Of the many food related poisonings and contaminations Europe had seen, this was by far the most systemic and potentially serious to human life (Philips 2001).

At the same time a series of initially distinct protests had been developing in Europe. An Anglo-Dutch controversy had taken place over the dumping of a redundant oil platform in the Atlantic Ocean. A team of Greenpeace activists had occupied the Brent Spar, owned by Royal Dutch Shell. To support this protest in continental Europe, a boycott of Shell petrol stations took place (Dickson and McCulloch 1996). Eventually the company backed down, embarrassing both the British and Dutch governments, both of which had insisted on the right of the company to follow its plans (Tsoukas 1999).[2] For the first time cross-border protests had halted the activities of a European multinational company. At the same time, in a series of unrelated protests over a road building programme in the UK, protesters had developed a new repertoire of protest forms and activities, drawing on the examples of Earth First! in the US and their Australian counterparts. Although unable to stop those roads already in progress, they ensured that when the government changed in 1997 the programme was stopped (Wall 1999). Equally importantly, it created a group of people committed to high intensity and high profile forms of protest who were not just skilled but also available for future protests (Seel et al 2000).

For many years activists have travelled and been able to diffuse their ideas in person as well as through correspondence and books, but during these protests the internet became a tool for the first time. Not only could the arguments for and against a case be diffused globally but also the methods and techniques of protest could be shared. Websites, email digests of news coverage and even web cams conveyed protests to anyone with internet access (Kahn and Kellner 2004). Although this was only the latest development in the planetary scope of the ambitions of the organic movement, it added an urgency and immediacy that had not been present before. It facilitated a new planetary wide mobilization that was conscious of its scope and scale in real time. Protesters domesticating some international issues in part created the mobilization, ecological questions from North America were transplanted into the UK, and French cheese makers shared concerns about peasant farmers. By bringing topics across borders, and mixing the local and global in novel ways, the protest gathered allies and purchase in public debates.

Several commentators have observed the way in which the technology in new crops also helped shape the protests. The technologies deployed in these crops had allowed particular traits to be inserted into the DNA of the plant; some of these traits had come from other crops and others from other genera of life. For example, toxins from a bacillus were inserted into a plant with the result that the toxin would be released when the plant was bitten by an insect – an inbuilt insecticide (Nottingham 1998). Companies such as Monsanto had focused on bringing these new seeds into a package of products and services that they would sell to the farmer. All of the benefits of these new technologies were realized on the farm, and for the companies that owned the patents on the plants and the associated chemicals. Any benefits to the end consumer were at best indirect. Although this has a certain degree of validity, it ignores the introduction of products such as the Flavour Savr tomato,[3] which was engineered to have a longer shelf life but found little market with consumers who were more disturbed by the fruit's failure to ripen rather than being impressed by its longevity.

This coincidence of factors did not occur to those in the global corporations who were looking to introduce the crops to Europe or it did not seem pertinent to them; after all regulators and consumers alike in North America had accepted them. The Food and Drug Administration (FDA) had accepted in 1992 that the crops were 'substantially equivalent' to conventional ones and no ill effects had been observed amongst the millions of people who had consumed them since (Nottingham 1998). Widespread protests about the way in which crops gained governmental approval seemed unlikely, as most people outside those departments or businesses involved were not aware of them. The advisers and consultants to these businesses had badly miscalculated, as several of these huge corporations were to be battered and broken on a wave of popular opposition.

Roots of the opposition

The first opposition to GM crops was launched as soon as scientific field trials of the plants were held in the early 1980s. Jeremy Rifkin took legal action against the trials of strawberry plants, which included genes that protected against freezing taken from fish (Rifkin 1998).[4] After this brief flurry of activity, the central area of opposition to the development of genetic technologies as a whole developed in Germany. As Rao makes clear, there was a sustained campaign in Germany by a relatively small, but determined, group of protesters against the involvement of a range of companies in the emergent technology (Rao 2009). This opposition forced German companies to develop genetic products in the more permissive environment of the US, so raising the costs of development. Through an alliance with the Green Party in Germany, the protesters worked to create uncertainty about the regulation and future commercial prospects of the products in Germany. This made these businesses more reluctant to invest and their staff less willing to be involved in the development process. It also raised with the wider public the ethical and regulatory issues around these new technologies. The roots of the opposition to GM technologies and their products are deeper and more sustained than many realized during the protests against GM crops.

The opening of the organic movement's revolt against GM technologies was sounded at the Lady Eve Balfour Memorial Lecture in the autumn of 1997. Jonathan Porritt warned of the danger of the 'Gene Dictators', equating democracy with the freedom to choose not to have these new technologies (Porritt 1998). Porritt had been a founder member of the British Green Party, and the director of Friends of the Earth in the 1980s. He had played an important role in brokering contacts between the environmental movement and the organic movement, after conflicts during the late 1980s. Although little noted outside the organic movement, Porritt's role as an adviser would mean that he would be able to lever far greater publicity in the new year through his friend the Prince of Wales. The naked protests at the meetings of the Food and Agriculture Organization (FAO),[5] the polite lobbying of legislatures and the pamphlets of campaigning groups could not garner the publicity that a simple statement from the future king of the UK could.

A controversial figure long before the death of his first wife, Lady Diana Spencer, Charles, Prince of Wales had become an organic farmer in the late 1980s and in his position as the world's most famous organic farmer he weighed into the debate (Charles, Prince of Wales and Clover 1993; Dimbleby 1994). Charles was close to the leadership of the Soil Association and at that time was a patron[6] of it. In an article in the *Daily Telegraph* he argued that, 'I happen to believe that this genetic modification takes mankind into the realms that belong to God, and to God alone' (Charles, Prince of Wales 1998). He proceeded to outline the discursive package of the opponents to GM food: the long-term effects on human health, increased use of agro-chemicals, genetic drift, insect resistance to pesticides, past agricultural disasters, inability to recall the product and the question of food supply. He topped this off with questions about the necessity of the technology and the freedom of choice of consumers. At the end of the year the Prince's new website, in a move symbolic of the whole campaign, started to host a debate on GM foods. Charles was able to launch a planetary campaign through global media publicity. The Soil Association had begun to campaign on the freedom of choice for consumers and the theoretical threats caused by GM technologies, but events were soon to outstrip their tactics.

To be accepted in the European Union, the new crops had to pass basic evaluations to ensure their compliance with the regulations, with each member state undertaking its own trials. These trials were widely distributed throughout the UK, often in proximity to organic farms. Concerns had been raised that the pollen from these crops would cross-fertilize with organic crops, causing a form of genetic pollution and, under organic certification rules, strip the crop of its organic status. One such crop was in the county of Devon, England, right next to one of the most high profile of the country's organic vegetable growers, Guy Watson (Gibbs 1999). Watson headed up a cooperative of growers in that area who franchised their distribution through a box scheme called Riverford (see Chapter 6). The GM trial was taking place in a concentration of organic farms, amongst which were some of those with the highest public profile. Watson took the matter to a judicial review, a tactic that the Soil Association had used successfully in the past to contest controversial technical decisions. After legal manoeuvring, the High Court gave Watson a judgment on 15 July 1999, that the government had been wrong to allow the trial to go ahead but as it was now planted it could remain. This Solomonic judgment signalled the end of the Soil Association's existing repertoire of protests. Watson was pictured in the local press warning that he feared that frustrated people might take 'direct action' against the GM crop. Sure enough on 3 August, the field was 'decontaminated' by a local group, who placed the plants in biohazard bags and waited to be arrested (Reed 2002).

The organic movement had found that it had to be assisted in this dispute by those with more experience of organizing protests that went beyond the law. Although broadly the environmental movement and the organic one had been in sympathy for nearly 30 years, they had not always agreed or acted in concert. It was only in the mid-1990s that groups such as Friends of the Earth came to recognize the value of organic agriculture (Lamb 1996). The organizations of the organic movement had no experience of involvement with or management of the protests that were beginning to break out. This did not mean that the organic movement was passive, it moved to a position where it was providing both the arguments against, and the counter example to, GM technologies. Their broader strategy of using consumerism to bear witness to the need for change in agriculture was twisted to give people wider options to take part in the protests.

Pollen and its promiscuity with other crops, particularly organic crops, became a way in which the conflict around GM crops could be dramatized and realized, particularly on a small, crowded island like the UK. Scientific disputes in the US, played out in the scientific journal *Nature*, raised the question of the damage done by GM crops on non-target species such as the monarch butterfly. Despite the very different ecology of the UK to that of the US, it provided a clear example of the possible dangers of GM crops. The Soil Association commissioned a report from the UK's National Pollen Unit, which concluded that maize produced up to 150lbs of pollen an acre, which at most could travel 180 kilometres and remain viable for up to nine days. Drawing its horns in for the conclusions, the report found: 'Overall it is clear that maize pollen spreads far beyond the 200m cited in several reports as being an acceptable separation distance to prevent cross-pollination' (Soil Association 1999). Genetic pollution from pollen could become an all-pervasive problem if the GM crops became a commercial reality, and organic agriculture might end. Although couched in careful tones, the report was the describing the doomsday scenario for organic agriculture. In this discourse the problem of GM was couched in scientific terms and restricted to discussions of agriculture, but protesters were already getting arrested in an attempt to provoke a test case.

During this period GM products that had appeared on the shelves of British supermarkets were being removed, as consumers actively avoided them and supermarkets were faced with the possibility of a boycott. All the supermarkets had recently seen the impact of the recent boycott of Shell as well as the impact of BSE on beef sales and were not going to risk their market share. A cocktail of food scare and boycott could be highly toxic to a retailer. At the same time organic was being promoted as a safe choice, free of GM and the taint of BSE;[7] it was being promoted as 'food you can trust'. With GM appearing throughout the food supply chain, the major European supermarkets, led by the French chain Carrefour, sought to purge their own products of GM products (*Daily Mail* 1999).[8] The market that the likes of Monsanto had hoped to capture was already drying up, as retailers acted in response to, and in anticipation of, their customers.

The Monsanto Files

At this time there were also threats and the fear of legal action by the biotech companies, particularly Monsanto. Authors such as Andrew Rowell had analysed the use of strategic lawsuits to stop environmental protesters in North America. These were often conducted alongside the use of public relations companies to mimic the community organizing of NGOs in a sustained effort to reverse the impact of the environmental movement (Rowell 1996; Beder 1998). It appeared that the same measures would be replicated in Europe. An example of this was when Monsanto threatened to take legal measures against all of those in possession of a handbook detailing how to destroy crops.[9] Into this fray stepped the editors of *The Ecologist*, relishing the possibility of legal action. His nephew Zac had joined Teddy Goldsmith, and they had compiled what they described as *The Monsanto Files*. *The Ecologist's* normal printers gave them a huge publicity coup by deciding that the contents were so likely to lead to legal action that they destroyed the first print run to ensure that they would not be a party to it. Once the Goldsmiths found a printer, the major retailers refused to distribute or sell it for the same reason as the printers. The edition eventually became the best-selling edition ever of the magazine, with global sales of 400,000 copies and no suit from Monsanto was forthcoming.

As the first article, the Prince of Wales Daily allowed his article from *The Telegraph* to be reprinted. The edition linked Monsanto to the defoliant, agent orange, that had been contaminated with PCBs, and which in the 1960s had been widely dropped over Vietnam during the war. It continued onto the techniques of strategic lawsuits used against activists by the corporation, the dangers of genetically modified bovine growth hormones, again made by Monsanto, onto the 'terminator technologies' that would limit the viability of GM seeds and, for a finale, the edition outlined the 'revolving door' relationship between government and executives in Monsanto. It also attacked the tactics of the PR company appointed by Monsanto, as well as its arguments and listed the companies using GM products. As always with *The Ecologist*, the solutions put forward were less industrial technology, more tradition and more organization on the small scale. Organic agriculture was lauded as the way forward, whilst the threat to organic standards in the US was noted. Andrew Kimbrell argued in the magazine:

> *Monsanto and other agribusiness conglomerates are seeing the birth of a powerful new competitor for consumers in the United States and Europe, organic food production* (Kimbrell 1998, p296).

Just inside the front flap of the magazine was an advert for the newly opened 'Planet Organic' supermarket in London and the first book advertised on the back page was the latest reprint of Howard's *Agricultural Testament*. *The Ecologist* was being consistent in its message and playing an important role in positioning the organic movement. Monsanto played the part of the blundering opponent to the full, but it was the early identification and characterization of the corporation as 'the enemy' that was central to the mobilization. *The Monsanto Files* played an important role in positioning the organic movement within the wider environmental movement, both in terms of organization and also its ability to argue.[10]

Action against the trials

The British direct action protests were building on an established legal defence in cases of non-violent direct action, that the sincerity of the protesters meant that they could escape the most severe sanctions of the law. As the protests continued, the level of provocation, particularly towards Monsanto, increased as the damage mounted and the viability of the trials became threatened (Burkenham 1999). Although conducted carefully and seriously, the protests incorporated elements of pantomime that sought to defuse any possibility of violence but also mocked those sponsoring the trials. The most serious attempt to generate a legal test case came when the head of Greenpeace in the UK, Peter Melchett, destroyed a trial crop near his family estate in Norfolk, England. Melchett and his collaborators did not just gather the plants by hand but used machinery to destroy the crop. The reason he struck was evidence from the Swiss authorities that they were receiving seeds from the US contaminated with GM material; the possibility of a non-GM future was declining (BBC Online 1999). Given his prominence and the publicity surrounding his jailing, it was clear that Melchett intended that his trial should be a chance to rehearse the problems of GM crops in a very public setting.

While this dispute continued, the British government was scrambling to save the trial process and its own policies. After nearly two decades of Conservative Party rule characterized by the political programme of Mrs Thatcher, the New Labour government of Tony Blair had been elected on a landslide. Blair had made a number of very pro-GM food statements, although it was apparent that the Cabinet was split over the issue (Cook 2004). The Labour Party had a long history of viewing new technologies as a path to social progress, although during its extended period in opposition several of members of Parliament had publicly sided with the organic movement. Critics of the protesters saw them as motivated, if not by anti-scientific views, then by misinformed ones and demanded that the policy was governed by science. This position was undermined by some of the government's own advisory agencies, particularly English Nature, which had the remit of protecting the English countryside, being against the early introduction of GM crops.

Inevitably in a situation as conflicted as the one that emerged in the UK during 1999, elected politicians seek a compromise, but the one that the government ministers responsible found on this occasion was remarkable. Michael Meacher and Geoff Rooker in consultation with the industry bodies, set in place a series of scientific trials that would prove the environmental consequences for the flora and fauna surrounding the crops. Independent researchers would set up pilot plots in 1999, which would then move to full-scale trials through until the autumn of 2002 (Vidal 1999). The findings would be reported in 2003 after being peer-reviewed by reputable and august scientific bodies. For the first time anywhere the full consequences of these crops would be validated through a scientific process that was both academically rigorous but also held in the full glare of public scrutiny. A scientific committee drawn from representatives of the industry, academia and NGOs would oversee the whole process. In the meantime there would be a moratorium on the commercial planting of GM crops. If Meacher and Rooker thought this would end the controversy, they were to be proved wrong (Reed 2006).

The French connection

Whilst the bailed Peter Melchett was waiting his moment in court for causing criminal damage, the flamboyant José Bové led a group from the Confédération Paysanne as they dismantled a McDonald's franchise in Millau in Southern France on 12

August 1999 (Bové 2001). The protest was ostensibly about the use by McDonald's of beef that had been produced using growth hormones. It was also symbolic of a wider attempt to publicize the claims of the Confédération in defence of the traditions of European agriculture against the depredations of multinational capitalism. Bové was a radical in the 1970s who, through supporting sheep farmers in a protest against the army, had become one himself, although his flock did not stop him from pursuing radical causes, including time on Greenpeace's *Rainbow Warrior* and working with anarchist groups. Bové's protest carefully linked a transatlantic trade dispute with wider questions of food safety, European culture, food sovereignty and globalization.

In February 1998 the World Trade Organization (WTO) had given the EU 15 months to rescind the ban it has placed on importing beef from the US. The EU opposed the widespread use of growth hormones in the US, a practice banned in the EU on safety grounds. When that period had expired and the EU had taken no action, the US, within the rules of the WTO, was able to retaliate by placing 100 per cent surcharges on a selection of 100 European food products. Amongst this basket of goods was Roquefort cheese, the key product of Bové's region and the spark for his protest. The picture of a grinning Bové, handcuffed and surrounded by gendarmes ,became a symbol of his success. Unlike the UK where criminal damage as part of a deliberate and sincere protest would see protesters walk free, Bové was facing gaol time, a prospect he appeared to relish. Like Melchett, he was seeking publicity; he had been courting arrest by destroying GM crops for several months but a McD's hit the spot, and, as the corporation might have said – he was lovin' it.

The battle for Seattle

The events during the meeting of the WTO in Seattle in November and December 1999 took many commentators by surprise as a coalition of trades unions, debt activists, peasant farmers, anarchists and those opposed to the North American Free Trade Agreement (NAFTA) took to the streets. This was the manifestation of a movement that had been growing throughout the 1990s, particularly in the Americas, that came to public attention as it battled the police in Seattle during five days from November into December 1999. They represented a section of thought that was generally more radical than the main thrust of the organic movement had become. The centrality that they placed on agriculture, the defence of rural life and opposition to the dominance of multinational corporations spoke to the radicalism of the early 1970s in the organic movement. If not immediate allies, each movement could recognize aspects of their ideas in the programmes of the other, and events would provide the early stages of cooperation.

The streets of Seattle became the fulcrum of a debate between those opposed to global trade and a police force determined to ensure that the meeting of the WTO went ahead. Thousands of protesters and activists from across the planet had made their way to the city, where some were determined to make their voices heard by the delegates and others to ensure that the meeting did not take place. One of the striking features of this ensemble of protesters was the breadth and diversity of the issues that they were representing and to a degree their solutions: traditional trades unionists, Earth First!ers, those focused on a single species, French farmers, Mexican peasants, the Sierra Club and anarchists found their way to the gathering, united only in their opposition to what they saw as the continuing neo liberal thrust of globalization represented by the WTO.

Who ever was responsible for the violence, the heavy handed police or protesters determined to break the law, the streets of Seattle became the cauldron for a series of street battles. The police deployed baton rounds, armoured vehicles and tear gas in a futile effort to ensure that the delegates arrived at the conference centre. The WTO was not able to open its meeting on 30 November 1999, resulting in the deployment of the National Guard and the imposition of a curfew to get the delegates through. Although only a minority of the protesters were involved in the confrontations, many of them receiving injuries as well as detention and occasionally legal sanctions, they captured the media's coverage and the sympathy of many of the other protesters in Seattle and beyond. Amongst those taking part in the protests, the counter-conference and debates in the city were, it has been claimed, farmers from over 80 countries (Bové and Durfour 2001, pxi). Food and rural issues were pushed to the forefront of global protests.

Amongst the tear gas and bravado of the protesters in Seattle the figure of José Bové provided an important link to an emerging global movement of peasants that connects with the critique launched by the organic movement but also differs from it. The centre of this movement is a critique of the green revolution and the project of 'developmentalism', as part of the globalization of capitalism. As discussed in Chapter 5, the organic movement had long criticized the technology of the green revolution and then, during the early 1970s, the global trade in food. During the 1980s Vandana Shiva became a doyen of much of the environmental movement for her book *The Violence of the Green Revolution* in which she examined the impact of the green revolution on the Indian province of the Punjab (Shiva 1991). Shiva argued that not only did the technologies of fertilizers, seeds and irrigation damage the environment, but they also exacerbated economic and ethnic tensions, often leading to violence. In a direct reference to the work of Albert Howard, Shiva saw the solution lying in a return to the organic farming tradition of India and an agriculture focused on maintaining a stable peasant society.

Shiva's work was part of a broader backlash against the concept of 'development' as increasingly being pushed by Western nations, multinational corporations and global development institutions such as the World Bank and International Monetary Fund (IMF). This form of development was seen as being a new version of the imperialism of the late 19th and early 20th centuries (see Chapter 1) whereby export-led agriculture undermines local communities and peasants are removed from the land to make way for agri-businesses trading commodities. Whilst Shiva's version of this is framed within some forms of romanticism or nostalgia, other parts of the trend framed it within a more pragmatic programme.

By the end of the 1990s this movement had brought together a global range of actors: small farmers from Europe such as Bové, farmers' unions from India and peasant movements in Central and South America. These latter groups drew direct links between global politics, inequalities in their own societies and questions of food and farming. The most well known example through this period was the Movimento dos Trabalhadores Rurais Sem Terra (MST), the landless rural workers movement of Brazil. They focus on issues of land rights, occupying uncultivated land, growing basic foodstuffs and building alliances with the urban poor. Through organizing in parallel to various global meetings on food and development, a series of alliances was formed between social movements, NGOs and other groups articulating an alternative vision of global relations and food production. In 1996 at the Forum on Food Security held in parallel to the World Food Summit, this new group launched the concept of 'food sovereignty'.

Food sovereignty is designed as a direct rejection of the global market in food commodities and can be best explained in four parts. The first is that people have 'the right to safe, adequate and nutritious food and healthy water' which 'is a fundamental human right of individuals and groups'. This includes supplying national markets, and for people to produce food for their families and communities. These are seen as social and cultural rights that are as important as the civic rights usually discussed in rights talk. Next is local control of natural resources, 'Peoples should have equitable access to and control over land, water and genetic resources necessary to maintain their livelihoods in a sustainable manner', rather than facing them being privatized or patented. Third is that the future of farming is not based on chemicals or genetic manipulation but a 'commitment to prioritising agro-ecology as the mainstream sustainable and appropriate production system for food and farming, livestock raising and fisheries'. Whilst this is not certified organic farming, it is a commitment to forms of agriculture and food production that are recognizably part of the broader thrust of the organic movement. The fourth is that this programme is not about isolationism or autarky but around retaining localized or more localized control; a 'commitment to promoting an equitable and fair trade system that is a positive force for development and does not detract from the realisation of any human right' (Campesina 2000).

In 2000 two events happened that brought some organization and visibility to this new movement. The first was the formation of the 'International Planning Committee Food Sovereignty' that sought to coordinate the range of organizations involved in this campaign across the globe. The second was a meeting of Via Campesina and Food First Information and Action (FIAN) in Honduras; in a declaration typical of this movement they looked forward to the 'Global Week of Struggle for Agrarian Reform' in December. FIAN is a network of activists 'fighting hunger through human rights' by responding to abuses of the right to food through lobbying, case work and monitoring. Via Campesina is an 'international movement of peasants, small- and medium-sized producers, landless, rural women, indigenous people, rural youth and agricultural workers' promoting sustainable agricultural production through family or peasant farming in harmony with local traditions. In a joint declaration they opposed GM technologies and the increasing globalization of agriculture and the economy:

We oppose to the production and commercialization of genetically modified organisms and to the privatization and patenting of living organism genetic resources. We reject these policies, that constitute severe threats to alimentary sovereignty of our families and our people. We are convinced of the fact that Agrarian Reform is a basic element to democratize land and economy. (Campesina 2000)

The struggle for agrarian reform, the opposition to GM foods and technologies, for this movement was part of a broader contestation of global capitalism in its neo-liberal form.

This movement is not concerned with a return to a past of a rural idyll or championing the status quo. Rather, as Philip McMichael argues, 'Instead of defending a world lost, transnational movements such as Via Campesina advocate a world to gain – a world beyond the catastrophe market regime' (McMichael 2008, p210). Raj Patel in *Stuffed and Starved* points out that the movement's focus on women's rights 'goes for the jugular in many rural societies, opening the door to a profound social change' (Patel 2007, p302). Yet, as Patel argues, there is a difference between the agro-ecology espoused by this movement and organic, as 'organic is now, however, an industry'. He continues by equating organic with not using pesticides but keeping the rest of the infrastructure of the food industry. Patel's alternative of growing as well as cooking your own, is perhaps at best misjudged but it is certainly a romantic one and based on a straw figure of 'industrial organics', that does not appreciate the rich variety of the movement.

Bové turned up at the Seattle protests to hand out Roquefort cheese outside McDonald's and to denounce the role of the WTO in further disempowering farmers across the world. Bové was explicit in drawing parallels between the impacts of the WTO on small and traditional farmers in the North and the fate of peasants in the South. The criticisms of the green revolution that the organic movement had done so much to foster and develop had been taken up and carried in a new direction even if that was a path that few in the organic movement were prepared to recognize or follow immediately.

GM on trial

Whilst the protests diffused globally and connected with agendas far more radical than the organic movement had been associated with before, the field scale evaluations of GM crops continued in the UK. Although far less dramatic and without the photogenic appeal of the protests that had become part of the global controversy, they did hold a particular position in the planetary debate as the only publicly sponsored societal debate based on experimental science. It became evident as the trials continued that it was only a generally bad tempered truce rather than a meeting of minds. Many of the groups opposed to GM trials saw the crops as too hazardous to be conducted outdoors, so the Soil Association, Greenpeace and Friends of the Earth had no formal place in the process. The direct action activists kept hunting down the crop trials and destroying them as their locations, including map references, were publicly listed. In 2000 it became clear that alongside the field scale evaluations the government had authorized a series of secret trials about which ministers in the same department had not been informed (Lean 2000). During the pressure of the 2001 general election it became apparent that many of the proponents of GM crops were not in total agreement with the process either.

Greenpeace and the Soil Association suspected that the crop trials were placed deliberately in proximity to organic farms to force action by certification bodies against organic farmers. Of the 104 large-scale trial sites, 30 were near organic farms whether this was skulduggery by the GM companies is not proven but it speaks to the mistrust between the parties that it could be imagined. One trial crop in May 2001 nearly wrecked the whole process, as the committee supervising the process was involved in a very public disagreement. The cause of this conflict was the discovery that Aventis had plans to plant a trial crop a few miles from the Henry Doubleday Research Association's (HDRA) demonstration gardens, where, as well as domestic gardening, trials of organic horticulture were being held. The Soil Association point- ed to the threat of GM pollen to the trials and experiments, including some funded by the government. Patrick Holden, the director of the Soil Association, deliberately put this in a global context: 'This is the worst example so far in a programme of insidious pollution of the world's food crops by the GM industry' (Reed 2006, p46). Michael Meacher, the government minister responsible, described the location as 'provocative' and the press published a leaked email from the chair of the committee supervising the trials describing concerns about the crop's siting as 'political' and concerned with 'public relations'.

The dispute continued to escalate as the Supply Chain Initiative on Modified Agricultural Crop (SCIMAC),[11] the trade body representing the GM crop compan- ies, insisted that it had the legal right to plant the crop, and Aventis had instructed it to be planted, in direct contradiction to a request from the Minister. SCIMAC argued that the minister's request did not raise any objections that met the condi- tions under which they could stop the planting. The HDRA replied that they were preparing to take the matter to law and Patrick Holden continued to build the rhet- orical pressure:

This is the forces of darkness deliberately trying to wreck organic agriculture by growing GM crops next to a centre of organic excellence. It is particularly scandalous that a body that has a vested interest in destroying organic agricul- ture should behave in this way. (Reed 2006, p48)

In a foot-in-mouth incident, a spokesman for SCIMAC claimed that; 'Just because this is a Labour marginal seat we are not giving into pressure. Mr Meacher is wanting to say, "I stopped the trial vote for us".' The whole process looked as if it was going to collapse into a series of court cases and a controversy that would destroy the process.

David Toke, in his analysis of the dispute, points out that the entire process was saved through the unlikely intervention of the Royal Society for the Protection of Birds (RSPB) (Toke 2002a, 2002b). The RSPB were sitting on the panel overseeing the trials; as one of the largest membership bodies in the UK, with over a million members, it has a well funded and respected staff of conservation policy lobbyists and scientists. The RSPB's Director of Conservation accused SCIMAC of 'bungling' the trials, seeking the conflict with organic farmers and threatened to withdraw from the process unless the trial plot was stopped. Although this threat could be interpreted as an act of solidarity with the organic movement, it was also self-interested on the RSPB's part, as they did not want to be associated with taking part in a discredited process. SCIMAC stopped the planting at that location and the crop trial process continued the search for scientific evidence, whilst around it the opposing groups viewed it as an opportunity to conduct their conflict through other means.

In anticipation of the findings of the field scale evaluations (FSEs) the British government fostered a wide ranging public debate called 'GM Nation', which began in June 2003. A report to the prime minister on the economic potential of the crops, as well as a review of the science surrounding them was complemented by an exercise to encourage public debates across the nations of the UK. Facilitation packs were made available for those wishing to organize debates in their communities, whilst a market research company conducted debates with a representative cross-section of the population. The initial goals of the project were dashed when it was revealed that the results of the FSEs would be delayed until the autumn, beyond the period scheduled for the debate. The exercise concluded that a majority of people were uneasy about commercialization of the technology and that the more they knew about the topic the harder their opposition became. The report from the market researchers listed the most common arguments in order of importance; first was the risk of cross-contamination: 'People argue that in a small island with mixed farming, co-existence of GM and organic is impossible'. The second reason was that it would destroy freedom of choice, 'People who use the arguments about contamination, or freedom of choice, make no distinction between different GM crops: they regard them all as incompatible with organic and other possibilities for farming and food'. In third place was the argument that GM crops were a risk to the environment. This extensive debate demonstrated that wide sections of the British public had accepted many of the central arguments of the organic movement and its allies (AEBC 2003).

The GM Nation debates overshadowed the findings of the FSE, as it demonstrated that the depth and breadth of public opposition went beyond the narrow agronomic questions the evaluations had been set up to answer. The findings of the FSE were very even handed, as it found that the different crops had both advantages and disadvantages to the surrounding wildlife. Oilseed rape and sugar beet crops resistant to herbicides had deleterious consequences for wildlife, whilst the maize had a beneficial effect, although one official remarked, off the record, that the trials had demonstrated that the wider impact of maize was so large that it brought into question growing it all.[12] The final twist was that the report on the findings of the whole process recommended that those who undertook to grow GM crops took out insurance that would cover them without admitting any criminal liability – an admission that the crops would not be free of controversy and that legal challenges could even be successful. The lobby group 'Farm' quickly published a survey of the major underwriters, including a damning quote from one of the insurance companies spokespersons: 'If a farmer approached us with any kind of insurance policy relating to a farm associated with GM we would have to refuse their application whatever the kind of insurance applied for'(Cropchoice 2003). Monsanto, which had taken the brunt of the public opprobrium after aggressively promoting the technology, did not wait to hear the verdict of the whole process, closing its facilities in the UK and scaling back its operation in Europe on the day the FSE results were announced (Connor 2003).

Learning from the mobilization

The mobilization against the introduction of GM technologies was the first time that the organic movement rose as a planetary whole in a manner that focused explicitly on legislation and the politics of food. In doing so it exposed its limitations, as the tactics of working within the rules, that had been used to advance the movement for 40 years, quickly became exhausted. Only the interventions of allies using new forms of protests created the political pressures that allowed it to succeed to the extent that it did. Other parts of the environmental movement that often used ideas that the organic movement had developed in the 1970s brought a new repertoire of protest into the organic movement.

Whilst the marketing of organic food and the development of organic farming propelled the movement into the orbit of agri-business and the state, the mobilization against GM crops pushed it in another direction. Because of the structure of the protests, the organic sector was a direct beneficiary of the protests, but much of what was mobilized against was associated with the sector. The organic movement came into contact with radicals in a way that it had not done since the 1930s, their programme being focused on the structure of agriculture, the rules of global trade and the dominance of their domestic economies by corporations.

If we take Naomi Klein's *No Logo* as a representation of many of the arguments of the anti-globalization protesters and their agenda, then clearly the existing organic sector meets many of their demands. As Heath and Potter point out, for all the radical arguments against global trade and corporations the positive response is to create a 'No Sweat' logo (Heath and Potter 2006). The insistence on national or local autonomy suggests that a large part of the movement was about protest against present developments rather than having any programme of change. Even those sympathetic to the protests were wary of the emphasis on national autonomy as cutting across solidarity between different people (Cockburn et al 2000).

Despite the sound and the fury unleashed in Europe, the movement was far less successful outside it, with many key areas supplying the European market, such as Brazil, making wide use of GM crops.[13] The organic movement and its allies demonstrated to the multiple retailers in Europe, and those who supply them, that they are capable of organizing significant boycotts but their influence on governments was far less effective, in part because of global trade agreements. This typifies the limitations of a cultural movement in that it can mobilize the amorphous and loose, create generalized concern or ideas, but specific actions are more difficult. The mobilization also brought into question the strategy of the movement as it focused attention on demands for swift and sweeping changes that went far beyond the specific technologies but also on the style and form of agriculture across the world. As seen in the previous chapter, the market-led strategy is incremental – one field, one stomach at a time – whilst through collective action and government intervention wider vistas were being opened.

Yet for the corporations it served as a powerful warning and also a lesson as to the rising power of the wider environmental movement. After a period in the early 1990s, when it appeared that through a combination of aggressive legal actions, 'astro-turf' organizations – whereby public relations companies organized seemingly genuine civic organizations – and slick advertising, that the environmental movement was on the back foot, through finding new allies, trades' unionists or the organic movement, the environmental movement found new areas in which it could press its case and in turn its allies found novel ways to protest.

This mobilization for the organic movement was also a rejoinder of its long running opposition to, and contestation of, the green revolution. Many of those organizations, and in some cases people, who had been involved in the implementation of the green revolution saw GM technologies as the next step. Unlike the late 1940s and early 1950s, the advocates of the next or doubly green revolution could no longer claim national security; communism had been defeated and the new threats to global security such as al Qaeda were not dependent on rural insurrections. The state had no clear agenda for backing these technologies immediately as they had over 50 years previously. The arguments of the environmental movement had also had an impact in that questions about the effect of these technologies on the people of developing countries were viewed seriously. Secondly, as the work of economists such as Amartya Sen has demonstrated, hunger was and is mostly about access to food rather than absolute shortage. The corporations could play on the economic advantages of having a thriving biotech sector, but this was not the same imperative as the first time.

It is important to note that the most successful area of the planet in resisting the introduction of GM technologies, with only a few GM plants gaining legal approval and some countries having no commercial GM crops present, is the European Union. This should not be used as a demonstration of the power of those opposed to GM crops, as, for example, GM crops are used in European animal rations but the GM soya is grown in Brazil. European consumers and the multiple retailers have been instrumental in making the global introduction of GM plants contested, reflecting in part their economic power, but also that the model of the EU does still reflect the broadly social and Christian democratic politics of the continent. Outside Europe that process has been contested, but the state has not been so likely to heed the movements and protesters. As Ian Scoones makes clear, in some countries GM crops were smuggled in for farmers to make use of them before they were properly regulated and little public debate took place before the protests appeared in Europe (Scoones 2008). It was the example of the European protests that encouraged those elsewhere. The repertoire of crop trashing was emulated, the arguments diffused and shared. Although it is possible to view some of these protests as self-interested protection by anxious wealthy consumers, there was also solidarity developed with the poor over the exploitation of farmers by multinational companies.

As Hugh Campbell has argued in several instances, organic farming was able to insert itself as the binary opposite to GM food (Campbell 2004) and, recast as agroecology, it reaches even further through the groups opposing GM technologies. The debate about GM food frequently became about a much wider range of issues than the technology itself, as Scoones argues of anti-GM mobilizations in India, South Africa and Brazil:

> It is increasingly bound up with a much wider debate new relations between contemporary capitalism and society: issues of sovereignty, inequality, rights, justice and so on – an attempt, perhaps, to create an alternative

'grand narrative', one counter-posed to the mainstream neoliberal worldview (Scoones 2008, p340).

As the links between the environmental movement, that for agrarian reform and the organic movement demonstrate, the connections between these issues can be fairly weak and ambivalent. The attempt to talk about the world's trading system, the right to food and inequality in a new way was far beyond the boycott of the major retailers but it was closer to those groups destroying GM crops across Europe.

The energies and tensions created by this mobilization came to combine with those produced by the developments in the organic market to create new pressures in the movement and permutations within the discourse. Examples from those remote from the neoliberal order, such as Cuba, started to combine with radical prescriptions for living with and adapting to climate change. The organic movement is poised to make a series of profound changes, which is the subject of the next chapter.

Just as the legal actions of the Soil Association reached their limitations in the fields of south Devon and the movement needed the assistance of those with other forms of protest, the boundaries of some of the organic movement were extended by this mobilization. The contradiction of trying to argue for a fundamental change in agriculture and supplying a multiple retailer became increasingly obvious during this period, just as sales of organic food surged because consumers saw it as a refuge from the new technology. At the same time activists in the South may have eschewed industrialized organics, but their farming methods would be recognized by Howard, Balfour and Steiner. The tensions and contradictions within the organic movement would be harder to hold together after this mobilization than they had been before.

Notes

1   The local police unit that specializes in dealing with extremism, subversion and threats to public order.

2   Later Greenpeace admitted that its calculations of the contaminants in the platform had been wrong by a factor of ten, but by then the protests were over and the platform was being dismantled in a Norwegian fjord.

3   Tomato paste concentrate was also marketed in the UK, but this was not from Flavour Svr tomatoes but a different genetically modified variety.

4   Rifkin is a policy activist who has a particular interest in the impact of technology on public policy; based in the US, he is influential within the EU. See http://www.foet.org/JeremyRifkin.htm.

5   The form of protest is often very culturally specific; naked protests seem more popular in Germany, whilst silly costumes are favoured in the UK and more direct forms in France.

6   Patrons are major donors to the Soil Association.

7   The contaminated feed that led to BSE was banned under organic rules, so no animal born of organic parents developed BSE. Some animals born outside the organic system but brought onto organic farms did. Therefore the UK movement was always careful in its public claims about organic and BSE.

8   Whilst WalMart is the largest global supermarket, two European groups are generally seen as being the second and third largest: Carrefour and Tesco respectively.

9   For a time it appeared that I would be included in this group, as I had a copy as part of a study into radical environmentalism, an opportunity that the University's legal team appeared eager to savour.

10 As a link to those early years, Bob Waller was still a member of the advisory board at this time.

11 See www.scimac.org.uk.

12 Personal communications.

13 Soya grown in Brazil is of great importance to animal and dairy production in Europe.

8

Peak Organics?

In 2008 full-blown organic consumerism began a terminal decline; it crawled to the end of a supermarket aisle gasping. Out of its ending the beginnings of a new period of the organic movement emerged and a new set of disputes entered the organic discourse. The ostensible reason for the failure of consumerism as the way of advancing the movement was the recession ushered in by the economic crisis, but as earlier chapters have shown, there was always ambivalence towards the strategy within the movement. As journalists tried to search for a shorthand for the 'excesses' of the boom years, organic food became a new cliché (Jowit 2008; Scmit 2008). The reasons for this change were deeper than opinion pieces and *vox pop* in daily newspaper and TV reports. The landscape of food and farming was undergoing a rapid and profound transformation; some of the flows causing this change stem from the economic crisis but others from tensions within the movement and pressures on the environment.

The economic crash, with at times the whole edifice of advanced capitalism teetering, barely disguised the global environmental perils running in parallel to the misdemeanours of the global speculators. Oil prices were increasingly volatile, reflecting not only the activities of speculators but also discussions of 'peak oil', that in the future it was going to become radically less available. A crisis in energy was rising to meet the challenges presented by climate change, as the planet began to show clear evidence of warming. These 'crunches' on resource use for many in the organic movement represented an opportunity, as the limits of nature were being found and so a reordering based on those demands was becoming imminent.

At the same time a slow burning discussion about food had arisen, partly born from the activities of the movement, and one that had become entangled in discussions about social justice, a debate the organic movement found itself ill-equipped to engage in. Celebrities combining their status as cook and activist began to pose questions about food culture that clashed explicitly with some of the marketing used around organic food, whilst some of those in the organic food chain realized that parts of that marketing had damaged organics, and sought to re-promote their organic products. The organic movement's strategy was reaching its limits just as the environmental and social boundaries it had long anticipated appeared to be taking a hold on global policy discussions.

In response to these strategic limits and shifts in the planetary context, the movement is beginning to alter; these changes have yet to coalesce into structures and forms that are clear-cut, but the tendencies are apparent. This chapter explores them in turn, under the overarching argument that the organic movement has hit a hiatus that it is struggling to overcome. To advance further, to become and achieve what it has promised in the past, it needs to find a politics, but in doing so it will have to sacrifice some of the advantages it has gained through being a cultural movement. The choices are stark, in that it can become just another farming technology that can be adopted without social change, it can develop into a prop to the marketing schemes of agri-business or it can grow to be involved in politics in a new way. None of these options is without cost, the question is about who pays this cost and when. It is this debate that much of the movement is currently engaged within various arenas and forms. The result of the debate will guide the shape of the organic movement into the future.

Anthony Giddens argues in his book on climate change that the challenge of the green movement has ended and is about to be absorbed by the mainstream because green values can be separated from green theories of agency. For example the green concern for direct democracy has little to do with the preservation of ecosystem through conservation measures. He argues that green ideas will become part of a wider politics and that some will be more widely adopted, for example questioning the importance of economic growth in already wealthy countries. For Giddens it is how we view the relationship between humans and nature that needs to be changed:

> *We must also disavow any remaining forms of mystical reverence for nature, including the more limited versions which shift the centre of values away from human beings to the earth itself – tackling global warming has nothing to do with saving the earth, which will survive whatever we do.* (Giddens 2009, p56)

Giddens' account of the green movement is superficial, but his two wider points are well founded: that global warming forces us to reconsider the relationship between nature and humans, and this requires a new politics beyond the currently established discourses. Throughout this chapter these are constant themes that are profoundly altering the debate about the future of humanity, and the ecosystems that now depend on it.

Within this transition to a new phase of the organic movement are three interconnected flows that challenge much of what has underpinned the movement until now. The evidence presented in this chapter is aimed to represent these trends. First are the debates about the resource crunch, how that is being used in the movement and by its allies. What initially appears to be an ideal opportunity for the movement falters over the implications of organics in these forms. Second, through a discussion of the role of celebrity chefs and their social activism, the limits of cultural politics are considered and explored. Third, a redefinition of the global definitions of organic food and farming is used as a way of considering the relationship between markets and the social goals of the movement. Finally, a suggestion is made as to the three paths the movement may have open to it.

Summiting the peaks

> *The most frightening aspect of the end of the oil age relates to the production and distribution of food. At present the populations of developed countries, especially those who live in urban conurbations, are totally dependent on 'just-in-time' distribution of food from a totally oil and gas dependent industry. If energy gets much more scarce and expensive and this precipitates trade conflict or war, it is easy to imagine a situation where cities literally run out of food. The possible consequences of the resulting chaos are disturbing to contemplate* (Holden 2007).

> *Modern agriculture has charted a course for disaster. Our soils and water resources, and our weakened food crops, will fail us just as energy depletion makes it increasingly difficult to make up for these deficiencies through artificial means. The fossil fuel-based agriculture that allowed our population to climb so far above carrying capacity in the last century will soon falter most cruelly. There is, however, a chance to transform ourselves into a more sustainable and equitable civilization* (Pfeiffer 2006, p82).

Throughout the 1970s the environmental movement predicted the end of a range of natural resources and the reckoning that would bring to industrial society. As *The Blueprint for Survival* made clear (*Ecologist* 1973, see Chapter 5), this was imminent and active planning for the change was required. The end of the first decade of the 21st century saw renewed discussion of resource exhaustion; oil, water and phosphates were all forecast to be peaking in extraction, with less remaining than had been used. At the same time anthropogenic climate change was moving from the margins of discussion to the main thrust of many government policies. Into this discussion the organic movement and its advocates in the wider environmental movement inserted arguments about a post-oil, low carbon world.

As can be seen from the two lengthy quotes above, this new turn in the discourse of organic food and agriculture repeats features of the older arguments but with new elements. In Holden's quote there is pessimism about the ability of the contemporary system to radically change. Mainstream economic theory would argue that, as oil becomes more expensive, the greater would be the incentive to innovate and produce non-oil alternatives. For the environmental advocates of the peak scenario, the contemporary system does not have enough time or the inclination to make such changes. The only resource available is a response from the grassroots, of ordinary people at a local level making the transition to a post-oil world, anticipating the problems that peak resources will bring. Mass and popular mobilization will stave off the chaos that global capitalism, without its lifeblood of oil, will wreak on people and the planet. Pessimism about the contemporary trajectory of global society is matched by unbridled optimism in the power of the popular masses. The environmentalists become the agents of global order, whilst capitalism represents chaos. Pfeiffer seems to follow much the same argument of pessimism but adds a new dimension, that of equity.

There are two exemplars of this transition to facing a world with less or no oil; the first is the experience of Cuba after its links to the Soviet bloc were severed in the early 1990s and the Transition Town movement in the UK. Each looks to different forms of popular mobilization to realize change.

Cuba

The end of the Cold War had a number of unexpected consequences for the global organic movement. For some it heralded an end to the logic of the green revolution, in that agricultural productivity ceased to be a bulwark against communism so the price of food could be allowed to rise and national self-sufficiency was no longer a strategic goal for any government. In the froth and jubilation by some at the triumph of capitalism, consumer preferences could be allowed to sway agriculture in a way that it had not before, hence organic farming could be allowed its head. In a curious parallel of these arguments, Cuba found itself cut adrift from its Soviet sponsors. After nearly 40 years of fraternal preferential exchange of sugar for oil and oil products, the communist regime found itself without the oil and still under trade blockade by its neighbour in the US. Cuba embarked on a crash programme of adopting organic farming and other experiments, as its oil economy dried up and the country entered a 'special period' of trying to feed itself.

Outside war, few countries can have experienced an economic collapse of such proportions, with the gross domestic product falling by nearly 50 per cent in the years between 1989 and 1993, from $19.3 billion to $10 billion. Imports of fuel fell by 70 per cent, fertilizers by 75 per cent, pesticides by 60 per cent and animal feed concentrates by 70 per cent. Through its agreements with the communist bloc countries, Cuba had been able to import nearly two-thirds of its food and with the collapse of its European partners faced the possibility of having to feed itself. To add to the complexity of these problems, the main agricultural export – sugar – was also its main chance to earn foreign currency. For all the gains of the revolution in education and scientific achievement, Cuba remained a country reliant on a commodity crop exported onto the world market, like so many of its neighbours in the global South.

The effort to turn Cuba into an island economy based on organic farming started in 1992 with the formation of a group of educators and researchers from the Ministry of Higher Education and the Agrarian University of Havana. As well as working to change the internal structures of Cuban agriculture to facilitate the low energy and intensive techniques of organic agriculture, the group worked to promote the production of organic cash crops – coffee, sugar and tobacco, urban agriculture in the cities and the reintroduction of animal tillage into Cuban farming systems. The distinctive element of this drive to national self-sufficiency was that it drew on the global contacts and supporters of not just Cuba but also the organic movement. Of those attending the first National Conference on Organic Agriculture in 1993, 40 people came from abroad, with 100 from Cuba. By the time of the third conference in 1997 there were 400 delegates, 180 from abroad and 240 Cubans. The extension projects set up by the group to popularize organic methods gained support from the United Nations Development Programme (UNDP) as well as Oxfam America, Bread for the World (Germany) and HIVOS from The Netherlands. By 1999 the group had been recognized by the Cuban state system as the Organic Farming Group (GAO), and later that year it won the Right Livelihood Award from the Swedish Parliament. In seven short years the group has achieved not only results in promoting organic agriculture in Cuba but also state and international status (Funes et al 2002).

Cuba as a society presented many aspects that were favourable to this widespread adoption in that as a government system it is both authoritarian and reliant on mobilization of the population. Therefore the GAO was able to use the mass media, television programmes as well as magazines, and rely on a population with a tradition of working in community groups, in a situation where organic methods represented a necessity rather than a conviction. The global organic movement was able to provide the technical assistance that allowed the group to develop systems of production that suited the needs of the Cuban population and the agro-ecology of the island (Wright 2009).

Many in the organic movement, particularly those in North America, use the example of Cuba as the exemplar of a society living in a post-oil, low carbon manner. Yet the example of Cuba is an ambivalent one, in that is has been achieved by a one-party dictatorship that has long practised forms of popular mobilization. It is hard therefore to argue that the Cuban people have chosen this form of agriculture; rather, it has become the option that has been presented to them with few alternatives. The Cuban system is also underpinned by food imports, which are not organic, from its North American neighbours. Therefore the Cuban experience demonstrates that organic food can be adopted as a technical system with few social or cultural implications. This is the mirror image of many of the critiques of the role that organics plays in the food system dominated by the multiple retailers. The dictatorship of the Castro brothers has no more changed its values than the CEOs of WalMart or Tesco have changed theirs. In a mirror image of the 1930s, parts of the organic movement have come to admire a dictatorship using a popular mobilization to conserve the planet. Both communists and capitalists have been able to strip away the wider implications of organic agriculture, as argued for by many of its proponents, and implement it as a technical system of agriculture. This questions whether there is a culture of agriculture that has broader consequences unless it is embedded in a political programme.

Transition Towns

The transition initiatives currently in progress in the UK and beyond represent the most promising way of engaging people and communities to take the far-reaching actions that are required to mitigate the effects of Peak Oil and Climate Change. (Brangwyn and Hopkins 2008, p4)

The arguments about Peak Oil have sparked the mobilization of a new network or movement looking to address a transition to the end of a carbon based economy. This movement calls itself Transition Towns (TT) and is based on the premise that peak oil has either occurred or is imminent, so immediate measures need to be taken to prepare for the end of an oil-based society. As part of this process, the movement calls for each community to create an Energy Descent Action Plan (EDAP) to put into place the measures needed to ensure that it could make this transition. To date this movement has been active across 85 communities in the UK, as well as in various locations in the US, Australia, New Zealand, Germany, Chile, The Netherlands, Italy and Japan. Alongside this roll call, the movement claims that 600 other communities are debating subscribing to the Transition Town process, an impressive rate of growth for a network founded in 2005, from the town of Totnes in Devon, England.

The aim of the TT process is to engage local communities in responding to the end of an oil-based society with creativity and optimism, through looking to make their area more self-sufficient and self-reliant. By establishing a series of groupings the community creates its own EDAP, through mapping their resources, 'visioning' their community in 15–20 years in the future, creating milestones, engaging with the local government's plans for the area and then finalizing the area's EDAP. The informal leader of the movement, Rob Hopkins, is clear that this is a deliberate process of creating culture change and reversing the common prejudice against the rural. Hopkins uses models of change drawn from the psychology of addiction as a guide, apparently viewing oil usage as a literal dependency. The movement argues that the process of change needs to be an inclusive one, not based on creating conflicts or opposing factions but an all-embracing process of community change.

Many elements suggested as part of the EDAP are familiar to most environmental or green programmes for the past 30 years, with a call for a relocalized economy, local currencies, time sharing, high density eco-efficient housing, mass public transport and now Cuban style integration of housing with horticulture and more broadly low carbon organic farming. In some of their arguments they refer to the environmental thinking of the early 1970s that focused on an imminent resource 'crunch' (see Chapter 6). Hopkins in particular uses phrases drawn directly from the writings of Rudolf Steiner; whether deliberately or not is unclear. The most direct source of the arguments is permaculture, a form of environmental planning based on a green philosophy. Hopkins has taught permaculture in Eire, and has sought to expand its systems thinking beyond the individual homestead to towns and even small cities. In doing so it draws in part on the networks and history of local action, of nuclear free and GM free zones, of the Steiner schools, local food and currency experiments, but reframes these within the urgency of peak oil. As Hopkins argues 'how to use peak oil as a tool for empowerment rather than leaving people feeling helpless. This part of the exploration is about how to actually facilitate change, and the dynamics of cultural transformation' (Hopkins 2009).

The TT movement shares the weaknesses of the green movement identified by Giddens, in that it assumes a relationship between its aims and the means it identifies as being necessary. By looking only to local solutions, it neglects the power of the central state to act and in the British context uses one of the weakest of arms of government to create change. In insisting that everyone can be included it also poses that these profound social changes are ones that cannot be opposed; rather than political difference it is supposing a moral hegemony. It echoes *The Blueprint for Survival* in a communitarian mobilization that might or might not prove to be entirely democratic. It also has an environmental determinism whereby resource limits have to be accepted as the motive force behind social change, rather than those changes being negotiated. A nature of limits replaces the central authority of the neoliberal state, the market, with little room for discussion.

The ends of nature

During the 1940s, and even beyond that period, Eve Balfour could write and talk of *Mother Earth*, of a nature that was bigger than humanity. This metaphor has been increasingly replaced by 'the planet' or 'the Earth'. The change is important as it moves from a nature that nurtures and is wise, through to a huge ecosystem that is enduring but vulnerable to human intervention. Yet, as climate change makes very clear, in many ways humanity is entering another mode of managing the planet's ecosystems. It may not be able to do so with precision but we are able to know through science the collective influence of human activities. No part of the planet's surface is untouched, even if indirectly, by human activities. Whilst no human foot may have trodden a particular sheet of ice in the vastness of the Antarctic ice sheet or the moss below a rainforest canopy, the changes to the climate induced by human activities mean that it is now within the limits of society. Bill McKibben wrote of this in his book *The End of Nature* in 1990, that nature as an entity or category outside humanity has ended (McKibben 1990).

This was apparent in the debate about GM technologies and the arguments against it launched by the organic movement. What many of the proponents of GM did not recognize, when accusing the organic movement of being anti-science, was that few people opposed the collection of the knowledge underpinning the new technologies; rather they opposed its deployment in those forms. Technology and its forms were being contested, not science. Nature as the mysterious force that humanity could not understand, a central part of the organic discourse in the 1930s, has been eclipsed. In that way the organic movement has become suffused with the insights of ecology and the environmental sciences. As the sociologist Ulrich Beck argues, all of contemporary society lives in the middle of an environmental crisis that we are only aware of because of the knowledge derived from science (Beck 1992).

The examples of Cuba and of the Transition Town movement demonstrate the difficulties for the organic movement of this changing relationship with nature, of which climate change is a sign. Cuba demonstrates that organic farming can just be a technology, not necessarily bringing a society or culture closer to nature. As a Marxist state, Cuba retains the materialist philosophy that views nature as matter that can be reshaped to human ends, as well as denying many of the rights of those people trapped within its grip. The Transition movement reverses this in positioning nature as implacable and the only answer being direct democracy, an environmentalist argument that denies the full social capacities of a society with deep democratic traditions. Neither realizes the full complexity of the organic movement as a form of democratic participation, itself nestled within a broader political system.

Climate change offers organic farming opportunities because it can be configured or reconfigured to not only be a low carbon form of agriculture but carbon-negative as it captures and holds carbon in the soil. In this it can connect with the redemptive technologies that the Howards offered in the 1930s and 1940s, an immediate answer to the environmental crisis of the moment. It provides a new impetus to measure all farming and food against the damage that it does to the wider environment. In these ways climate change is an affirmation of many of the central arguments of the organic movement. Yet the change in the status of nature is a profound challenge to all environmentalists, as there is no longer an exterior nature to which to return. Therefore all arguments about nature are societal debates; if nature and society could be separated previously, it is no longer possible to do so as directly.

From pukka to policy

Few public discussions of how people eat, what can be done to change it and the way in which the food system links with consumer health can be held in the UK without someone mentioning the celebrity chef Jamie Oliver. Oliver, known as 'The Naked Chef' to TV viewers and cookery book readers across the world, reinvented himself as a philanthropist and food activist with his documentary series *Jamie's School Dinners*. He sought to change the school catering provision in one London borough through overhauling the dinner menu. Part of the privatization drive of the 1980s had seen school meals turned over to profit-making enterprises. International catering companies took up these opportunities, quickly catering to the lowest common denominator of profit yielding fast foods desired by their junior customers. Oliver was appalled by the low quality, low preparation fare that the children were offered and set about attempting to provide food with more fibre, nutrients and greater skill needed in preparation. The resulting reality TV, with profane arguments, rebellious parents and students made for viewing so compelling that it began to mobilize a new coalition of parents and those working in education and public health, eventually leading to a change in government policy. Oliver's demonstration of how to make a processed poultry product involving a blender, hyper pink bird flesh and a class of horrified teenagers spread as a viral video.

Oliver had started as a champion of organic food and, although he had a major contract with the multiple retailer Sainsbury's, he had been scathing of supermarkets: 'For any chef, supermarkets are like a factory. I buy from specialist growers, organic suppliers and farmers' (Barton 2009). This earned him an equally public rebuke from the company chairman, Justin King. Oliver's original intention in *School Dinners* had been to get all of the food served to be organic but this proved to be far beyond the budgets that the education authorities could command. He returned in 2008 with *Jamie's Ministry of Food*. The opening programme was a searing polemic on the food and food habits of the British underclass, including a section where a young single mother living on benefits explained that she could not cook so spent most of her frugal budget on feeding herself and her daughter, leading them into a spiral of debt. The passage included images of her daughter eating slivers of a kebab from a polystyrene box whilst sat on the floor. Oliver's tearful rage was caught as he sat in his Range

Rover parked in the street outside. Later in the series Oliver visited the same woman, whom he had taught to cook basic meals, sitting at a table eating home cooked food.

Oliver and his team were deliberately invoking the wartime Ministry of Food that taught people to cook with the limited rations that they had. This was the high point of British collectivism, a time when the diet and health of the poor improved, alongside that of the rest of the population, after the state took control. As *School Dinners* had demonstrated, it was not just a question of supply but of skills and food culture. This led Oliver into a difficult collaboration with a woman who had very publicly undermined his efforts at healthy eating in the previous series. The reaction to Oliver's intervention was revealing of the tensions around food culture and class in the UK. Joanna Blythman, often an advocate of organic food and a scathing critic of British food culture, missed the point (Blythman 2004, 2006), accusing Oliver of blaming the victims: 'No, the people who really need to be grilled here and told to mend their ways are government ministers. I'd like to see them carpeted, not the teen mums with overdue electricity bills in their hands and kids at their ankles' (Blythman 2008). Whilst many might join in this fantasy of justice, it left unanswered Oliver's point that lack of skills to look after yourself makes the experience of poverty even worse than it might otherwise be. Felicity Lawrence, writing in *The Guardian*, argued that the programme was 'an indictment of the current political system as disturbing as any ideological tract. Food, and real people's experience of it, is still about class' (Lawrence 2008). In the early 21st century the documentary has become an ideological tract, the political pamphlet in video form and Oliver has levelled a powerful critique and solutions but offered a weak societal analysis.

Hollows comments in an early academic paper on how Oliver works on the boundaries of culture, in that he made it acceptable for young men to cook, blurring a traditional gender role in British life (Hollows 2003). In his TV documentaries he blurs the line between activism and documentary by pressing against the cultural meeting place of food culture and class in British society. The ambivalences of a multi-millionaire, who makes TV adverts for a supermarket, pointing out the stark social structural inequalities, erosion of healthy food habitats and the destruction of domestic economies by the market for food proved to be a potent mix. Jamie's genuine lament that he was 'fucking angry and I don't know what with but I'm fucking angry' exposes the problem for the organic movement. It should have been very clear to Oliver who the target of his anger should have been; he should have had a 'frame' for his anger provided by the organic movement. There was no clear script for his anger, for the social change that he had set about attempting to make. The answer to these problems was not the consumption of organic goods but a change to the food system and social structure of British society. Oliver had mobilized hundreds of thousands through his documentaries and yet the organic movement was unable to translate this into a broader mobilization. His documentaries were remarkable interventions in their brio and targeting of culturally awkward areas in British society. They are also remarkable in the inability of his allies to capitalize on the opportunities they were presented with. Oliver's journey charted the gap between premium organic brands and the transformation of societal attitudes to food, of the movement's inability to translate cultural influence into policy changes. He is just one example of a number of journalists and film-makers, Morgan Spurlock not least, who have made similar critiques but have not been able to translate them into a broader critique.

New definitions

Throughout the history of the movement, what organic means and how it can be defined has been a constant theme. In the immediate post-war period many people in the movement thought that science could play a role, whilst through much of the 1980s and 1990s it appeared that certification had provided a way of at least producing organic products. In 2005 the general assembly of IFOAM decided that it was necessary to create a new definition of organic, one that was not based on what organic farming is not, or existing legal definitions. A task force created for the purpose was given the project of surveying public definitions, those in previous IFOAM documents and definitions used in regulations. Through this review process they were to reach a synthesis that would define organic agriculture in its broadest sense for the beginning of the 21st century (Luttikholt et al 2008). Various drafts were presented on the IFOAM website, and member organizations and the general public were invited to comment on the definitions and the iterations of the task force. A final draft was arrived at in December 2007, with the Standards Committee reviewing this draft in the following month. Their comment was to remove the word 'food' from the definition, not to focus solely on farming but to allow the definition to include textiles and cosmetics (IFOAM 2008a). In March 2008 the World Board of IFOAM adopted the new definition.

That a global movement that had spent nearly the whole of the previous century arguing a case should feel that it needed to reiterate its definition and so its purpose is indicative of the pressures the movement felt itself to be under. Through this reiteration the movement began the process of realigning its discourse:

> *Organic agriculture is a production system that sustains the health of soils, ecosystems and people. It relies on ecological processes, biodiversity and cycles adapted to local conditions, rather than the use of inputs with adverse effects. Organic agriculture combines tradition, innovation and science to benefit the shared environment and promote fair relationships and a good quality of life for all involved* (IFOAM 2008b).

The initial parts of the definition are consistent with the ideas of Balfour and Howard nearly 70 years before, but it is the second sentence that challenges aspects of the market as it has developed. It will be how 'fair relationships' are defined or operationalized that will determine the degree of change from present practices.

The definition was backed up by four principles: of health, ecology, fairness and care. Health and ecology reinforce the importance of agriculture that is specific to the ecosystems that it is lodged within but also that it must work to improve and enhance the health of that ecosystem and all those dependent on it. Rather than expressing total opposition to synthetic chemicals, the principle of health says that, 'it should avoid fertilisers, pesticides, animal drugs and food additives that may have adverse health effects'. The principle of ecology calls for those producing organically to reduce their resource consumption and to conserve resources. These principles challenge producers and consumers to create a form of agriculture that might not be part of the present practices of the multiple retailers.

The principles of fairness and care pick up the social agenda of organic farming laid down nearly 40 years before by writers and activists such as Fritz Schumacher and Barry Commoner. Fairness states that:

> *Organic Agriculture should build on relationships that ensure fairness with regard to the common environment and life opportunities. Fairness is characterized by equity, respect, justice and stewardship of the shared world, both among people and in their relations to other living beings* (IFOAM 2008c).

Later it is even more direct:

> *Fairness requires systems of production, distribution and trade that are open and equitable and account for real environmental and social costs* (IFOAM 2008c).

These injunctions are not just aimed at farmers but all of those working throughout the food chain.

The principle of care highlights the importance of the precautionary principle within organic agriculture and that science needs to be balanced with a respect for accumulated wisdom:

> *Organic Agriculture should be managed in a precautionary and responsible manner to protect the health and well-being of current and future generations and the environment* (IFOAM 2008c).

This care needs to be reflected in decisions made about the development of organic agriculture:

> *Decisions should reflect the values and needs of all who might be affected, through transparent and participatory processes* (IFOAM 2008c).

These principles are intended by IFOAM to guide the development of positions, programmes and standards, and to be used as a whole rather than selectively.

The principles both rearticulate the positions of the organic movement and at the same time state them for the first time in the contemporary context. Many in the organic movement would have identified the key elements of the principles but, as the Task Force's discussions make clear, many parts of the movement were working with definitions that were negative and excluded the principles of care and fairness. In setting out these principles IFOAM is both creating an aspiration that the movement will seek to realize them, and it is also potentially demarcating what may not be in the future considered as organic. Many operating in the organic market may not share this vision of organic production, and certainly it is not yet reflected in legal processes of certification. The principles are a reassertion of the importance of the social movement; the danger is that they may only serve to map the gap between the movement and the industry it has created.

Stalled by the critics?

This is not a niche market. It is a strategic response to unprecedented threats that our food systems are facing (Holden 2009).

As Patrick Holden, director of the Soil Association, was forced on public radio to re-state the strategy of the movement in 2009, it appeared that the progress of organics through consumerism was floundering. The press that had delighted in simplistically reporting the ever upward growth of the statistics of organic food sales were equally happy to do so when they pointed in the other direction. The more informed critics of the organic movement had already taken more nuanced positions. Michael Pollan broadly echoed the argument of Julie Guthman about the creeping industrialization of organics. Others looked more directly at the strategy of the movement rather than the logics of capitalism to detect the problems facing contemporary organics. Raj Patel, in his book *Stuffed and Starved*, lays bare the inequalities of the food system that, as the title suggests, leaves some underfed and others burdened by excess. Patel draws a distinction between supermarket organics and that which is outside that system. He acknowledges that organics is an improvement, but suggests that false choices are made between organic and non-organic farming, whereas an informed agro-ecology would produce both the social and ecological goods that are necessary. Patel is more interested in the social offer of the organic social movement, which he argues is eclipsed in the supermarkets, rendering the choice between organic and non-organic as the choice between 'Coke and Pepsi' (Patel 2007, p246). For Patel ethical consumerism is the 'honey trap' that lures us into thinking that the only way in which we can communicate in the food system is through the market place. It is a mistake that he argues makes us misunderstand the way in which social change can be created:

> It [Ethical Consumerism] makes us think that we are consumers in the great halls of democracy, which we can pluck off the shelves in shops. But we are not consumers of democracy. We are its proprietors. And democracy not merely where when we shop, but throughout our lives (Patel 2007, p312).

Patel sees an endless struggle of differences and that we need to act collectively for greater equality. The mechanisms for that change are collective movements that he links directly to the third call of the French revolution – equality.

As discussed earlier, Heath and Potter have a broader concern with cultural politics, arguing that it is founded on the idea that difference will create social change, and in doing so makes a graver error than just mistaking the effect as the cause; but that difference is the driving engine of consumer capitalism. That difference from the mainstream becomes hooked around to become the new cool product, as people compete to be cooler and more different: 'The struggle for status is replaced by the quest for cool, but the basic structure of competition remains unchanged' (Heath and Potter 2006, p330). Their argument about organic food is that: 'Organic food is yuppie food, in our view, because the extra cost buys nothing more than distinction and an unfounded sense of moral superiority' (Heath and Potter 2006, p347).

They contrast this with buying a hybrid car, which is genuinely ethical consumption, 'because the consumer voluntarily agrees to bear a greater portion of the social cost of his driving, beyond what the law requires' (Heath and Potter 2006, p346). They further claim that the organic movement is based on 'a popular appeal based upon false and unconscionable health claims' (Heath and Potter 2006, p347). Unlike many in the environmental movement, they argue that they are concerned about what businesses do rather than their scale.

Box 8.1 *Beyond organic*

Academic researchers began to talk of a post-organic group in the early years of this decade, as they found some small growers and farmers who had ended their organic certification. These growers found the cost and strictures of certification to be too burdensome so they negotiated with their customers – to whom they largely sold directly through box schemes or market stalls – that they would continue to farm organically without certification. In Europe organics was becoming over involved with the supermarkets and policy circuits of the EU, whilst in the US it was becoming too dominated by big business, according to these growers and farmers. This led to many discussions of the role of certification in creating a challenge to the food system and the potential of organics to create a wider social transformation (Moore 2005).

Obviously if organic food were to produce more than just distinction then Heath and Potter would have to agree that organics is ethical consumption, and they might want to look at the environmental and social research that suggests that it does (see Chapter 1). Their argument about difference and branding driving capitalism is well made regarding much of the branding around organics, a process that in turn seeks to incorporate nebulous concepts such as 'local' and 'natural'. Their solution differs from that of Patel, as they see the role of the state as being important and more effective than often suggested. They suggest restricting the role of competition in society through school uniforms, ending tax breaks on advertising and working to end antisocial forms of competition. The state would play a role on behalf of the collective or society in limiting the pursuit of status and in turn the aspects of capitalism that are driven by that hunt.

The sociologist Stuart Lockie has recently suggested that there is little contradiction between the figure of the responsible consumer of alternative food and the demands of a neoliberal market (Lockie 2009). Through careful analysis, Lockie shows how the consumer of alternative foods can become exactly the sort of citizen neoliberalism requires. Again questioning whether the consumer oriented approach of organics can achieve wider societal changes. The British environmental writer George Monbiot, in his book *Heat*, attacks the hypocrisy of most environmentalists; he cites a campaigner who spends their holidays snorkelling in the Pacific. Monbiot's argument is that environmentalists do not stand outside society, that they need to be restrained like everyone else. He argues that individualized restrictions, a strategy adopted for nearly 40 years, 'alone is a waste of time' (Monbiot 2007, pxxiv). Only the state can force us to do the right thing, which will require legislation to restrain and regulate all of us.

These authors are not unsympathetic to organics; although they have differing arguments and modes of analysis, they all agree on the importance of social movements and the state in the future of food. What they all assert is the importance of politics and looking beyond the market for solutions to the problems faced by those trying to effect change. Individual change is unlikely to be enough without concerted collective action, including that of the state alongside it. Also, all are emphasizing the importance of larger structures in society, be it an unfettered market or the competition to be cool associated with it. Within these critiques is politics, in that there are struggles for power, antagonisms of interest and conflicts of opinion that need to be reconciled within democratic systems but that should not be eliminated or ignored (Mouffe 2005). The frequent recourse, such as in the TT movement, that there is a single moral answer is the negation of that politics.

As the examples in this chapter have demonstrated, there are gaps within the contemporary organic movement. What ought to be an opportunity for the organic movement, in the coincidence of climate change and peak oil, is likely to be lost unless it can find a political framework to engage with it beyond well-worn calls to direct democracy or examples of the political 'other'. In place of the Hunza we are presented with the Cubans, living examples of an existing alternative but separated not just by space but also by political culture. The emphasis on individual behaviour and a lack of political critique has meant that the potential gains that could have been realized from, for example, the cultural politics of Jamie Oliver and others have been lost. Not knowing how to translate the critique into a policy framework has demonstrated the limits of the strategy of consumerism. The challenges of fairness, participation and transparency represented by the IFOAM definitions are hard to reconcile with the behaviour of the multiple retailers and some entrepreneurs within the organic movement. All of these critics, journalistic or academic, point to the need to forge a politics more encompassing than that of the limited expression of preferences through the shopping basket. So far the organic movement has not been able to respond to these calls.

The fourth phase and three directions

In Chapters 1 and 2 I set out an argument that the planetary organic movement has developed in four interlocking phases, which have informed the chapters up to this point. It is now time to consider the fourth phase. This phase began in the middle part of the last decade, building from the critiques and mobilization around GM food, as well as the growing realization of the urgency of environmental change. At present some of these changes are inchoate, but a clear trajectory can be seen.

Table 8.1 *The fourth phase of the organic movement*

Phase	1st phase	2nd phase 1945–1969	3rd phase 1971–2005	4th phase 2005–
Arguments	Soil science and rural communities	Based in soil science and agriculture	Environmentalism – pollution in the wider environment, contamination of food, conservation	Environmentalism, resource conservation, global warming, social justice, carbon footprints, access to food
Strategy	Discussion and writing	Scientific proof of superiority of organic farming and food	Practical experimentation to create viable system. Develop a market for organic produce	Citizenship, popular and community mobilization, state action
Tactics	Critical community	Research stations and demonstration farms/plots	Develop a practical farming system and low-level consumption. Develop institutions to support the market	Governmental lobbying, state support for organic farming, local and community organic gardening, community supported organic agriculture
Role of participants	Discussion	Financial support, gardening and discussion.	Consumerism and discussion	Citizenship, debate, gardening, cooking

This fourth phase has not yet become dominant; indeed at present it would appear to be largely confined to sections of the UK and US movement, partly because of the oppositional stance that these parts of the movement have adopted, and also because inherent within this phase of the movement is a critique of the neoliberal political order, which is less marked in some of the more social democratic European nations.

Box 8.2 *Fair trade and organics*

The fair trade movement rose quickly across Europe and the US in the 1990s, from its origins in religious groups wanting to demonstrate solidarity with the poor in the planetary South. A growing range of products from coffee through to fruit and wine have gained fair trade status, indicating that those producing the crop have received a fair rather than market price for the goods and that workers have been treated according to a set of agreed principles. For some observers this form of certification could be matched to that for organic production, resulting in a product that is both environmentally and socially certified. Indeed some products, particularly coffee are now certified as both fair trade and organic. This then opens the question as to why fair trade is only applied to products from the South of the planet. Why would people in the North of the planet want to trade fairly between themselves? To date no joint certification between organic and fair trade has been achieved, and fair trade remains a label for commodity products exported from the South to the North.

This chapter has presented a collage of contemporary influences on the organic movement from a celebrity raging in the confines of a Rotherham council estate, through to urban gardens of downtown Havana across to the graphs and charts pored over at global summits of those determining international resource policy. The connecting discourse between them has been that organic food and farming can play a transformative role in the relationship between food and society. As the other chapters in this book have demonstrated, the organic argument is not fixed, rather it changes and mutates over time. That is, part of these debates speaks to the success of its campaign to be included in the discussion about food provision, yet this success has exposed its weaknesses.

Within this fourth phase are three tendencies that are in tension with one another. Each of these could become dominant or hybrids could appear, as the proponents of each gather support or advantage.

1   *Organics as technics*: that the organic system can be stripped of its social aspirations, and reproduced as a technical system. It can be reworked to become less carbon intensive, even perhaps involved in carbon sequestration and can be adopted widely, independently of certification or any wider social agreements. This could see organic systems adopted very widely as part of the retooling of global agriculture, supported by an extensive group of agricultural experts but with a diminishment of the movement aspects. It would not have ambitions for broader cultural or social change but would achieve ecological and environmental targets.
2   *Organics as commodity*: that the system of organic certification becomes the platform around which organics remains part of the 'retail offer' to Western and other wealthy consumers. This may even see the merging of this form of certification

with 'fair trade' to create an 'ethical' option for consumers. Although retailing is currently suffering through the recession there is no reason to suppose that it will not return to the levels of prosperity seen before the economic crash. This would be accompanied by a boutique supply of organic products through box schemes, farmers' markets and online sales that would supply a cachet to the supermarket system. It would retain some of the movement determined to improve the food system through constant critique and counter example.

3   *Organics for equity*: that organics as a system of agriculture intertwines itself with social goals more closely and links accreditation as organic to social goals. This could aim for mass small-scale production, cooperative structures for production and retailing, and high environmental standards linked to systems of enterprise that maximize the social return from agriculture, with the state playing a role in fostering and supporting organic agriculture because it secures a wide range of public goods.

A re-reading of this book will reveal that these three tendencies have been present throughout the development of the organic movement. Many agriculturalists are shocked to discover that organic farming has social goals and certainly some urban activists struggle with the needs of food production beyond the allotment. All movements are alliances, held together by fellow feeling and narratives of solidarity that allow the collective to form, the sense of the 'we'. At a moment of transition such as the time we are currently in, that alliance is being renegotiated. The scathing tones taken by Patel that the choice between organic and non-organic is that of 'Coke or Pepsi' captures how those more associated with the arguments for equity can view those who look to the arguments around organics as a commodity.

The route through which these tensions can be debated, and a new consensus achieved, would appear to be the concept of 'citizenship', as it unites the three tendencies. For those who view organics as technics, as a system that can be implemented without regard to the social and cultural discourses it is interwoven with, citizenship is an important reminder that technology has social consequences. Shiva's critique of the green revolution in the Punjab is focused on the detrimental effects of implementing agricultural technology without regard for the social context it operates within. One of the strengths of the organic movement has been the critique of technology without a consideration of the social and cultural, particularly in the work of Commoner and Schumacher. It would mean that those using 'organic' methods in Cuba could not describe it as such because it is not developed in a democratic context.

The arguments against viewing organic solely as a brand or commodity have been well rehearsed in this chapter and citizenship demands that the consumer is considered more widely than someone who is solely making an economic choice. It demands that people are given choices about agriculture and food in other arenas than the market place. In doing so it requires, at the very least, a dialogue between those who produce and retail food and those who eat it. The consequences of this perspective range from seeing consumers as partners in the values of the brand, that it is not the plaything of the marketers, through to viewing support for environmental and social benefits of agriculture as being part of the role of the state and so reflected in the tax bill. In turn it may mean that farmers have to accept that others have a role in running their farms.

For those who want to see organics as equity, they may feel that they have already embraced citizenship, but many of them have not considered its full implications. Many, but not all, of the experiments in small-scale production, local marketing and ecological protection have been socially exclusive in that they have been largely the preserve of the educated and included, as well as not having to address the problem of feeding everyone. They have been backed by the knowledge that food will be available. Citizenship requires that feeding everyone is an imperative, that the social and moral exclusivity that has been attractive in many of the experiments cannot be continued. The question is not of realizing virtue in my neighbourhood or foodshed but through the wider community of which I am part, a very different proposition. Citizenship is explicitly political; it shifts the locus of debate from the largely moral domain of relationships with nature through to relationships with other people.

The direction that the movement is likely to take is hard to be sure of at this moment. As an example at the household level, my household, the veg box scheme that has been the cornerstone of our family diet remains defiantly wedded to providing organic vegetables at prices below those of the multiples and even non-organic produce. It is a cooperative of producers and the coordinating functions are held in a trust to ensure it continues to follow its founding ideals. But it also supplies a range of breads, pastas and meats that have a provenance as exclusive as their price, whilst nationally the organic industry is rallying to take the promotion of organic into its own hands rather than leave it to individual brands or multiples.

The movement's organizations try to position organics as a low carbon or even carbon capturing food system to weave it into the post-Copenhagen conference world of food. Former members of the Clinton administration write to *The Economist* to lambast them about their sympathetic coverage of Monsanto and the debate about GM foods continues to simmer. Michelle Obama may have an organic garden at the White House but the regulation of organics in the US appears to be deeply divided between movement organizations and corporations supplying the national market. There are tensions remaining between the market and the movement but at this moment it appears that the movement is reasserting its control.

As Chapter 6 argued, the organic movement was important in the foundation of the contemporary environmental movement in the late 1960s and early 1970s through the work of Commoner, Schumacher and Goldsmith. As Anthony Giddens has argued, environmentalism as politics has largely failed, and certainly in the case of organics it is the lack of a politics of organic food that is part of the transition the movement is experiencing. Returning to Commoner and Schumacher it is apparent that both looked to a broader societal project in which the citizen played a central role. Commoner looked to an active and engaged citizenship that stemmed from the American Revolution, upholding the highest ideals of civic life. Part of this was taking forward a respect for the Earth and agriculture but it was not the totality of the engagement required; citizenship was more than selecting produce, but was taking part in debate and discussion. Schumacher's thinking may have derived more from spiritual sources but it also engaged with the idea of a citizenship that, whilst accounting for nature, placed people in the centre of the discussion. Part of the transition to the fourth phase of the organic movement may be a remembering of these facets of the movement's own history.

Come so far, still so far to go

The organic movement has travelled a long way from the scattered networks of the late 1920s and early 1930s, when the Howards in their research station and those around Steiner started to reconsider the relationship between agricultural technology and healthy food. Those early networks where tens of people exchanged letters, experimented on small plots, lectured to small audiences and eventually published a steady stream of books are a long way from the global webs of products and people that are the contemporary movement. Few in the contemporary movement would think, as Albert Howard did, that a concentration camp was a suitable place for dietary experiments or subscribe to the combination of aristocracy and organic farming of the Kinship in Husbandry. For many people organic food has become the symbolic heart of a global cosmopolitanism, of a regard for provenance that suggests being culturally and culinarily informed. As events unfold around food security, and as the pioneers of the movement would have argued, food is far too important to be only about fashion and social distinction.

From these early networks of people who visited one another, exchanged letters and texts, the movement is now woven together by products, certification agencies, representative groups, legally guaranteed standards, websites, emails, magazines and books. Often these are affirmed by what is shared in the interaction between people, of belonging to a group or network of people who recognize one another as sharing the same interests and concerns. It is these concerns and interests that are beginning to shift, as the relationship between people and the planet is reasserted. In the beginning of the movement many thought that social inequality between people was part of the natural order, whilst increasingly those in the movement would argue that the only future for nature is in greater equality between people.

For many years the movement focused on demonstrating its ideas by echoing the way in which agricultural technologies were disseminated by putting ideas into practice and allowing people to witness the difference. It then shifted to allowing people to purchase that difference, to support the existing alternative through selecting it from a peer group of products. Yet as the planet warms, access to food falters for many and others realize the tenuous logistical threads that connect them to the fragile food chains of the global market, consumer preferences appear an increasingly brittle bridge to securing their food futures. It would seem that culture has its limits, even in a deeply democratic society, to achieving the forms and aims of a durable and healthy food system.

The organic movement was created through the cultural hybridity of Western scientists looking at the traditional practices of other farmers and growers. It would seem that it is perhaps time for the organic movement to learn from the experiences of peasants, the landless and marginal in the planetary South who increasingly talk not of the right to choice in a global market place but of rights to food security and sovereignty. The shape of the next phase of the organic movement will be formed by the play of social needs and the politics that it generates. Without the organic movement the dominant food system would be unchallenged, the health of the environment and humans far worse. The challenge for the organic movement now, if it is to achieve more than ameliorate the worst aspects of the system it opposes, is to put in place a political discussion about how to feed every person on the planet whilst safeguarding its future.

Bibliography

AEBC (2003) *GM Nation*. London, AEBC

Aitchtey, A (1995). 'The "Hole in the Bucket" in medical care the Peckham Experiment was not permitted to repair – by government negligence', *Contemporary Review*, vol 266, pp236–243

Aldington, R. (1976) *Selected Letters of D. H. Lawrence*. London, Penguin

Arvidsson, A. (2005) 'Brands – A critical perspective', *Journal of Consumer Culture*, vol 5, no 2, pp235–258

Aschermann, J., U. Hamm, S. Naspett and R. Zanoli, (2007) 'The organic market', in W. Lockeretz (ed.), *Organic Farming: An International History*. Wallingford, CABI, pp123–152

Baker, D. (1996) *Ideology of Obsession: A K Chesteron and British Fascism*. London, Tauris

Balfour, E. (1943) *The Living Soil*. London, Faber and Faber

Balfour, E. (1946) 'Our association', *Mother Earth*, 1

Balfour, E. (1966 (approx.)) Soil Association Council. Rolf Gardiner Archive

Balfour, E. (1975) *The Living Soil and the Haughley Experiment*. London, Faber and Faber

Barton, G. (2001) 'Sir Albert Howard and the forestry roots of the organic farming movement', *Agricultural History*, vol 75, no 2, pp168–187

Barton, L. (2009) 'The people's chef'. *The Guardian*. London

Bateman, D. and N. Lampkin (1986) 'Economic implications of a shift to organic agriculture in Britain', *Agricultural Administration*, vol 22, pp89–104

BBC Online (1999) 'Lord Melchett refused bail'. *BBC Online*. 27 July

Beamont, P. (1993) *Pesticides, Policies and People*. London, The Pesticides Trust

Beck, U. (1992) *Risk Society*. London, Sage

Beder, S. (1998) *Globalspin: The Corporate Assault on Environmentalism*. Vermont, Chelsea Green

Belasco, W. (2007) *Appetite for Change. How the Counterculture Took on the Food Industry*. Ithaca, Cornell University Press

Best, G. (ed.) (1972) *Water Springing from the Ground*. Fontwell Magna, The Springhead Trust

Blake, F. (1994) *The Organic Farming Handbook*. Ipswich, Crowood Press

Blythman, J. (2004) *Shopped –- The Shocking Power of British Supermarkets*. London, Fourth Estate

Blythman, J. (2006) *Bad Food Britain: How a Nation Lost its Appetite*. London, Fourth Estate

Blythman, J. (2008) 'Jamie Oliver is blaming the wrong people', *The Guardian*

Bové, J. (2001) 'A farmers' international?' *New Left Review*, vol 12, Nov–Dec, pp89–101

Bové, J. and F. Durfour (2001) *The World Is Not For Sale. Farmers Against Junk Food*. London, Verso

Bramwell, A. (1985) *Blood and Soil: Walther Darre and Hitler's Green Party*. London, Kensal Press

Bramwell, A. (1990) *Ecology in the Twentieth Century*. London, Yale University Press

Brander, M. (2003) *Eve Balfour, The Founder of the Soil Association and the Voice of the Organic Movement*. Haddington, East Lothian, The Gleneil Press

Brangwyn, B. and R. Hopkins (2008) 'Transition initiatives primer – Becoming a transition town, city, district, village, community or even island'. *Transition Town Network*. Totnes, Transition Town Network, version 26

Brown, L. (1998) *The Shoppers Guide to Organic Food*. London, Fourth Estate

Bryant, A. (1943) 'Kinship in husbandry'. Lord Northbourne, Arthur Bryant Archive,

Bryant, A. (1943/4) 'National book club and organic writers'. *The Collected Papers of Arthur Bryant*. J. Hudfield. London, Arthur Bryant Archive

Burkenham, O. (1999) 'Coming a cropper: the undoing of Monsanto's GM dream', *The Guardian*, 22 October

Busch, L. (2000) 'The moral economy of grades and standards', *Journal of Rural Studies*, vol 16, pp273–283

Campbell, D. (1966 (approx.)) 'Haughley Experiment'. Rolf Gardiner Archive

Campbell, H. (2004) 'Organics ascendent: Curious resistance to GM', in R. Hindmarsh and G. Lawrence (eds), *Recoding Nature: Critical Perspectives on Genetic Engineering*. Sydney, University of New South Wales Press

Campbell, H., C. McLeod and C. Rosin (2005) 'Auditing sustainability: The impact of EurepGAP on organic farming in New Zealand', in G. Holt and M. Reed (eds), *Sociological Perspectives of Organic Agriculture: From Pioneers to Policies*. Wallingford, CAB International

Campesina, V. (2000) Declaration international meeting landless in San Pedro Sula – Honduras, Via Campesina

Carson, R. (1991) *Silent Spring*. London, Penguin

Castells, M. (1996) *The Rise of the Network Society*. London, Blackwells

Castells, M. (1997) *The Power of Identity*. London, Blackwells

Cavett, D. (2007) 'When that guy died on my show'. *Talk Show. New York Times*. 2009: Blog entry at www.opinionator.blogs.nytimes.com/2007/05/03/when-that-guy-died-on-my-show/ (02/05/2009)

Charles, Prince of Wales (1998) 'Seeds of disaster', *Living Earth*, no 199, pp6–7

Charles, Prince of Wales and C. Clover (1993) *Highgrove. An Experiment in Organic Gardening and Farming*. London, Simon & Schuster

Chase, M. (1992) 'Rolf Gardiner: a cross-cultural case study' in B. J. Hake and S. Marriot (eds), *Adult Education Between Cultures*. Leeds, University of Leeds

Clarke (1982–1983) 'Letter', *Soil Association Quarterly Review*, Winter

Clarren, R (2005) 'Land of milk and honey'. Salon.com (10/02/2008)

Clunies-Ross, T. (1990) *The Politics of Organic Agriculture*. Bath, Bath University

Cockburn, A., J. St Clair et al (2000) *5 Days that Shook the World*. London, Verso

Commoner, B. (1966) *Science and Survival*. London, Viking Press

Commoner, B. (1971) *The Closing Circle: Nature, Man and Technology*. London, Knopf

Conford, P. (1988) *The Organic Tradition*. Dartington, Green Books

Conford, P. (1998) 'The book which began it all'. *Living Earth, Bristol, The Soil Association*

Conford, P. (2001) *The Origins of the Organic Movement*. Edinburgh, Floris Books

Conford, P. (2005) 'Robert Waller: Poet, BBC producer and editor of *Mother Earth. The Independent*,

Connor, S. (2003) 'Crops giant retreats from Europe ahead of GM report', *The Independent*, 16

October

Cook, G. (2004) *Genetically Modified Language*. London, Routledge

Cook, G., M. Reed et al (2007) *Research Report: The Discourse of Organic Food Promotion: Language, Intentions and Effects*. Milton Keynes, The Open University 15

Coombes, B. and H. Campbell (1998) 'Dependent reproduction of alternative modes of agriculture: Organic farming in New Zealand', *Sociologia Ruralis*, vol 38, no 2, pp127–145

Corley, H. (1957) *Organic Farming*. London, Faber and Faber

Coupland, P. (1998) 'The blackshirted utopians', *Journal of Contemporary History*, vol 33, no 2, pp255–272

Cropchoice (2003) 'UK organization finds no insurance for biotech crops'. *Cropchoice.com*. www.cropchoice.com/leadstry7e98.html?recid=2112 (07/10/2003)

Crossley, N. (1999) 'Working utopias and social movements: An investigation using case study materials from radical mental health movements in Britain', *Sociology*, vol 33, no 4, pp809–830

Crossley, N. (2002) *Making Sense of Social Movements*. Milton Keynes, The Open University Press

Daily Mail (1999) 'French supermarket chain Carrefour bans GM', *The Daily Mail*, 6 February

Davis, M. (1999) 'A world's end: Drought, famine and imperialism (1896–1902)', *Capitalism, Nature and Socialism*, vol 10, pp3–46 no 2

Davis, M. (2002) *Late Victorian Holocausts: El Nino Famines and the Making of the Third World*. London, Verso

Davis, M. (2006) *Planet of Slums*. London, Verso

Davis, M. (2007) *The Monster at our Door. The Global Threat of Avian Flu*. London, The New Press

Dealler, S. (1996) *Lethal Legacy: BSE – The Search for the Truth*. London, Bloomsbury

Della Porta, D. and M. Diani (1999) *Social Movements: An Introduction*. London, Blackwell

Della Porta, D. and M. Diani (2006) *Social Movements: An Introduction*, 2nd edn. London, Blackwell

Della Porta, D. and S. Tarrow (2005) 'Transnational processes and social activism: An introduction', in D. Della Porta and S. Tarrow (eds), *Transnational Protest and Global Activism*. London, Rowman and Littlefield, pp1–17

Dickson, L. and A. McCulloch (1996) 'Shell, the Brent Spar and Greenpeace: A doomed tryst?' *Environmental Politics*, vol 24, no 1, pp122–129

Dimbleby, J. (1994) *The Prince of Wales*. London, Cape

Dudley, N. (1991) *The Soil Association Handbook*. London, Optima

DuPuis, E. M. and D. Goodman (2005) 'Should we go "home" to eat?: Toward a reflexive politics of localism', *Journal of Rural Studies*, vol 21, pp 359–371

Duram, L. A. (2005) *Good Growing. Why Organic Farming Works*. Lincoln, University of Nebraska

Ecologist (1973) *A Blueprint for Survival*. London, The Ecologist

Edgell, D. (1992) *The Order of Woodcraft Chivalry 1916–1949 as a New Age Alternative to the Boy Scouts*. New York, The Edwin Mellen Press

Egan, M. (2007) *Barry Commoner and the Science of Survival*. Cambridge, MA, MIT Press

Eyerman, R. and A. Jamison (1989) 'Environmental knowledge as an organisational weapon: The case of Greenpeace', *Social Science Information*, vol 28, no 1, pp99–119

Eyerman, R. and A. Jamison (1991). *Social Movements: A Cognitive Approach*. London, Polity Press

Fergusson, J. (2007) 'A menu for murder', *The Guardian*

FitzGerald, K. (1968) *Ahead of their Time: A Short History of the Farmer's Club*. London, Heinemann

Ford, B. (1996) *BSE: The Facts*. London, Corgi

Friends of Oswald Moseley (2001) Jorian Jenks. M. Reed personal communication

Fromartz, S. (2006) *Organic Inc. Natural Foods and How They Grew*. New York, Harcourt

Funes, F., L. Garcia, M. Bourque, N. Perez and P. Rosset (eds) (2002) *Sustainable Agriculture and Resistance*. New York, Food First Books

Gander, A. (2001) *Adrian Bell: Voice of the countryside*. London, Holm Oak Publishing

Geier, B. (2007) 'IFOAM and the History of the International Organic Movement' in W. Lockeretz (ed.), *Organic Farming: An International History*, Wallingford, CABI, pp175–187

Geiryn, T. (1999) *Cultural Boundaries of Science – Credibility on the Line*. Chicago, University of Chicago Press

Gibbs, G. (1999) 'Crown drops case against two "who wrecked GM crop"', *The Guardian*

Giddens, A. (2009) *The Politics of Climate Change*. London, Polity Press

Glassner, B. (2007) *The Gospel of Food. Everything You Think You Know About Food Is Wrong*. New York, HarperCollins

Greer, A. (2005) *Agricultural Policy in Europe*. Manchester, Manchester University Press

Griffiths, R. (1998) *Patriotism Betrayed*. London, Constable

Griffiths, T. and L. Robin (eds) (1997) *Ecology and Empire*. Keele, Keele University Press

Grove, R. (1996) *Green Imperialism*. Cambridge, Cambridge University Press

Guthman, J. (2004) *Agrarian Dreams. The paradox of organic farming in California*. Berkeley, University of California Press

Harrison, R. (1964) *Animal Machines*. London, Vincent Stuart Ltd

Harrison, R. (1970) *Can Britain Survive?* (Ed.) E. Goldsmith. London, Tom Stacey

Hassanein, N. and J. R. Kloppenburg (1995) 'Where the grass grows again: Knowledge expertise in the sustainable agriculture movement', *Rural Sociology*, vol 60, no 4, pp721–740

Heath, J. and A. Potter (2006) *Rebel Sell: How Counterculture Became Consumer Culture*. Chichester, Capstone

Heckman, J. (2006) 'A history of organic farming: Transitions from Sir Albert Howard's *War in the Soil* to USDA National Organic Program', *Renewable Agriculture and Food Systems*, vol 21, no 3, pp143–150

Heeks, E. (2007) 'Case history: Abel & Cole', in S. Wright and D. McCrea (eds), *The Handbook of Organic and Fairtrade Food Marketing*. London, Wiley-Blackwell, pp130–149

Heelas, P. (1996) *The New Age Movement*. London, Blackwell

Heinberg, R. (2007) 'The essential re-localisation of food production', in R. Hopkins and P. Holden (eds), *One Planet Agriculture. The Case for Action*. Bristol, The Soil Association, pp13–17

Hightower, J. (1976) *Eat Your Heart Out: How Food Profiteers Victimize the Consumer*. New York, Vintage

Holden, P. (2007) *One Planet Agriculture*. Bristol, The Soil Association

Holden, P. (2009) Interview. BBC Today Programme, 08/05/2009

Hollows, J. (2003) 'Oliver's Twist: Leisure, labour and domestic masculinity in *The Naked Chef*', *International Journal of Cultural Studies*, vol 6, no 2, pp229–248

Hopkins, R. (2009) 'Transition culture – About' www.transitionculture.org (03.05/2009)

Howard, A. (1940) *An Agricultural Testament*. Oxford, Oxford University Press

Howard, A. (1945) *Farming and Gardening for Health and Disease*. London, Faber and Faber

Howard, L. E. (1953) *Sir Albert Howard in India*. London, Faber and Faber

IFOAM (2008a) 'Considerations from the Task Force about the Definition of Organic Agriculture', Final Draft. Bonn, IFOAM

IFOAM (2008b) 'Definition of Organic Agriculture', www.ifoam.org/growing_organic/

defin-itions/doa/index.html, last accessed 10 June 2009

IFOAM (2008c) 'The Principles of Organic Agriculture', www.ifoam.org/about_ifoam/principles/index.html, last accessed 10 June 2009

IFOAM (2009) 'Global Organic Agriculture: Continued Growth', www.ifoam.org/press/press/2008/Global_Organic_Agriculture_Continued_Growth.php

Jackson, C. (1974) *J. I. Rodale. Apostle of Non-conformity*. New York, Pyramid Books

Jasper, J. M. (1998) 'The emotions of protest: Affective and reactive emotions in and around social movements', *Sociological Forum*, vol 13, no 1, pp397–424

Jenks, J. (1959) 'Editorial', *Mother Earth*, vol 13, pp133–134

Jenks, J. (1959) 'Editorial (1)', *Mother Earth*, vol 13, p10

Jordan, G. and W. Maloney (1997) *The Protest Business: Mobilizing Campaign Groups*. Manchester, Manchester University Press

Jowit, J. (2008) 'Shoppers lose their taste for organic food', *The Guardian*

Kahn, R. and D. Kellner (2004) 'New media and internet activism: from the "Battle of Seattle" to blogging', *New Media and Society*, vol 6, no 1, pp87–95

Kimbrell, A. (1998) 'Why biotechnology and high-tech agriculture cannot feed the world', *The Ecologist*, vol 28, pp294–298

King, F. (1927) *Farmers for Forty Centuries*. London, Jonathan Cape

Kirwan, J. (2004) 'Alternative strategies in the UK agro-food system: Interrogating the alterity of farmers' markets', *Sociologia Ruralis*, vol 44 no 4, pp395–415

Lamb, R. (1996) *Promising the Earth*. London, Routledge

Lampkin, N. (1990) *Organic Farming*. Norfolk, The Farmers Press

Lampkin, N., C. Foster, S. Padel, and P. Midmore (1999) *The Policy and Regulatory Environment for Organic Farming in Europe*. Hohenhiem, University of Hohenhiem.

Langman, M. (1989) 'A short history of the Soil Association', *Living Earth*, p21

Lawrence, F. (2006) *Not on the Label: What Really Goes into the Food on your Plate*. London, Fourth Estate

Lawrence, F. (2008) 'Britain on a plate', *The Guardian*

Lean, G. (2000) 'Revealed: Secret GM crop trials', *The Independent*

Lear, L. (1997) *Rachel Carson: The life of the Author of* Silent Spring, Allen Lane London

Leggett, J. (2006) *Half Gone. Oil, Gas, Hot Air and the Global Energy Crisis*. London, Portobello.

Lobley, M., M. Reed, and A. Butler (2005) *The Impact of Organic Farming on the Rural Economy in England*. London, Defra, p144

Lobley, M., A. Butler and M. Reed (2008) 'The contribution of organic farming to rural development', *Land Use Policy*, vol 26, no 3, pp723–735

Lockeretz, W. (2002) *Strategies for Organic Research*. UK Organic Research Conference, University of Wales, Aberystwyth

Lockeretz, W. (2007) 'What explains the rise of organic farming', in W. Lockeretz (ed.), *Organic Farming: An International History*. Wallingford, CABI, pp1–9

Lockie, S. (2009) 'Responsibility and agency within alternative food networks: Assembling the "citizen consumer"', *Agriculture and Human Values*, vol 20, pp193–201

Lockie, S., K. Lyons, G. Lawrence and K. Mummery (2002) 'Eating "green": Motivations behind organic food consumption in Australia', *Sociologia Ruralis*, vol 42 no 1, pp23–40

Lockie, S., K. Lyons, G. Lawrence and D. Halpin (2006) *Going Organic. Mobilizing Networks for Environmentally Responsible Food Production*. Wallingford, CABI

Ludovici, A. (1935) *Recovery: The Quest of Regenerative National Values*. London, The English Mistery

Luttikholt, L., K. Ageson, and R. Horwat (2008) *Definition of Organic Agriculture Report to the Task Force*. Bonn, IFOAM

Marcus, I. (1997) 'Sweets for the sweet: Saccharin, knowledge, and the contemporary regulatory nexus', *Journal of Policy History*, vol 9, no 1, pp33–47

Massingham, H. (ed.) (1944) *The Natural Order*. London, Dent

Masumoto, D. M. (1996) *Epitaph for a Peach: Four Seasons on my Family Farm*. New York, HarperOne

Matless, D. (1998) *Landscape and Englishness*. London, Reaktion Books

McAdam, D. (1986) 'Recruitment to high-risk activism: The case of freedom summer', *The American Journal of Sociology*, vol 92, no 1, pp 64–90

McAdam, D., S. Tarrow, and C. Tilly (2001) *Dynamics of Contention*. Cambridge, Cambridge University Press

McKay, G. (1996) *Senseless Acts of Beauty; Cultures of Resistance Since the Sixties*. London, Verso

McKay, G. (ed.) (1998) *DIY Culture: Party and Protest in Nineties Britain*. London, Verso

McKibben, B. (1990) *The End of Nature*. London, Penguin Books

McMichael, P. (2008) 'Peasants make their own history, but not just as they please ....', *Journal of Agrarian Change*, vol 8, no 2, pp205–228

Melucci, A. (1985) 'The symbolic challenge of contemporary movements', *Social Research*, vol 54, no 4, pp789–816

Melucci, A. (1989) *Nomads of the Present*. London, Hutchinson

Melucci, A. (1996a) *Challenging Codes. Collective Action in the Information Age*. London, Routledge

Melucci, A. (1996b) *The Playing Self: Person and Meaning in the Planetary Society*. Cambridge, Cambridge University Press

Melucci, A. and L. Avritzer (2000) 'Complexity, cultural pluralism and democracy: Collective action in the public space', *Social Science Information*, vol 34, no 4, pp507–527

Michelleti, M. and D. Stolle (2007) 'Mobilizing consumers to take responsibility for global social justice', *The Annals of the American Academy of Political and Social Science*, vol 611, no 1, pp157–175

Moloney, K. (2006) *Rethinking Public Relations*. London, Routledge

Monbiot, G. (2007) *Heat. How We Can Stop the Plane Burning*. London, Penguin

Moore, O. (2005) 'What farmers' markets say about the perpetually post-organic movement in Ireland', in G. Holt and M. Reed (eds), *Sociological Perspectives of Organic Agriculture: From Pioneers to Policies*. Wallingford, CAB International

Moore, O. (2006) 'Understanding postorganic fresh fruit and vegetable consumers at participatory farmers' markets in Ireland: Reflexivity, trust and social movements', *International Journal of Consumer Studies*, vol 30, no 5, pp416–427

Moore-Colyer, R. J. (2001a) 'Back to basics: Rolf Gardiner, H. J. Massingham and "A Kinship in Husbandry"', *Rural History*, vol 12, no 1, pp85–108

Moore-Colyer, R. J. (2001b) 'Rolf Gardiner, English patriot and the Council for the Church and Countryside', *Agricultural History Review*, vol 49, no 2, pp187–209

Moore-Colyer, R. J. (2004) 'Towards "Mother Earth": Jorian Jenks, organicism, the right and the British Union of Fascists', *Journal of Contemporary History*, vol 39, no 3, pp353–371

Moschitz, H. and M. Stolze (2007) *Policy Networks of Organic Farming in Europe*. Stuttgart,

University of Hohenhiem

Mouffe, C. (2005) *On the Political*. London, Routledge

Niggli, U. (2007) 'FiBL and organic research in Switzerland', in W. Lockeretz (ed.), *Organic Farming: An International History*. Wallingford, CABI, pp242–252

Nottingham, S. (1998) *Eat Your Genes*. London, Zed Books

Organic-Market.info. (2008) 'USA: organic ginger from China recalled', Organic-Market.info

Osborn, F. (1948) *Our Plundered Planet*. Boston, Little, Brown and Company

Patel, R. (2007) *Stuffed and Starved. Markets, Power and the Hidden Battle for the World's Food System*. London, Portobello

Payne, V. (1972) *A History of the Soil Association*. Manchester, University of Manchester

Pearse, I. (1979) *The Quality of Life*. Edinburgh, The Scottish Academic Press

Pearse, I. and A. Crocker (1943) *The Peckham Experiment*. London, Allen and Unwin

Pepper, D. (1984) *The Roots of Modern Environmentalism*. London, Routledge

Pepper, D. (2005) 'Utopianism and environmentalism', *Environmental Politics*, vol 14, no 1, pp3–22

Perkins (1997) *Geopolitics and the Green Revolution*. Oxford, Oxford University Press

Peters, S. (1979) 'Organic farmers celebrate organic research: A sociology of popular science', in H. Nowotny and H. Rose (eds), *Counter-Movements in the Sciences*. Dordrecnt, D. Reidel Publishing Company, vol III, pp251–275

Petrini, C. and G. Padovani (2006) *Slow Food Revolution. A New Culture for Eating and Living*. New York, Rizzoli

Pfeiffer, D. A. (2006) *Eating Fossil Fuels: Oil, Food and the Coming Crisis in Agriculture*, New York, New Society Publishers

Philips, Lord (2001) *The Inquiry into BSE and Variant CJD in the United Kingdom*. London, 3 volumes

Pollan, M. (2006) *The Omnivore's Dilemma*. London, Penguin

Polletta, F. and J. M. Jasper (2001) 'Collective identity and social movements', *Annual Review of Sociology*, vol 27, pp283–305

Porritt, J. (1998) 'The wisdom of wholeness', *Living Earth*, no 197, pp11–13.

Portsmouth, Lord (1941) *Alternative to Death*. London, Faber and Faber

Portsmouth, Lord (1965) *A Knot of Roots*. London, Geoffrey Bles

Pretty, J. (2002) *Agri-culture: Reconnecting People, Land and Nature*. London, Earthscan

Pretty, J., C. Brett, D. Gee, R. E. Hine, C. F. Mason, J. I. L. Morison, H. Raven, M. D. Rayment, and G. van der Bijl (2000) 'An assessment of the total external costs of UK agriculture', *Agricultural Systems*, vol 62, pp 113–136

Pretty, J., A. S. Ball, T. Lang and J. I. L Morison (2005) 'Farm costs and food miles: An assessment of the full cost of the UK weekly food basket', *Food Policy*, vol 30, pp1–20.

Purdue, D., J. Durrschmidt, P. Jowers and R. O'Doherty (1997) 'DIY culture and extended milieux: LETS, veggie boxes and festivals', *Sociological Review*, vol 4, no 2, pp645–667

Rao, H. (2009). *Market Rebels – How Activists Make or Break Radical Innovations*. Princeton, Princeton University Press

Reed, K. (1998) 'Whats in a label? Organic hash from the USDA kitchen', *Washington Post*

Reed, M. (2001) 'Fight the future!: How the contemporary campaigns of the UK organic movement have arisen from their composting of the past', *Sociologia Ruralis*, vol 41, no 1, pp131–146

Reed, M. (2002) 'Rebels from the crown down: The organic movement's revolt against agricultural biotechnology', *Science as Culture*, vol 11, no 4, pp481–504

Reed, M. (2004) *Rebels for the Soil: The Lonely Furrow of the Soil Association 1943–2000*. Bristol,

University of the West of England

Reed, M. (2006) 'Turf wars: The attempt of the organic movement to gain a veto in British agriculture', in G. Holt and M. Reed (eds), *Sociological Perspectives of Organic Agriculture*. Wallingford, CAB, pp37–56

Reed, M. (2009) 'For whom?: The governance of the British organic movement', *Food Policy*, vol 34, pp280–286

Reed, M., A. Butler and M. Lobley (2008) 'Growing sustainable communities – Understanding the social and economic footprints of organic farms', in C. Farnworth (ed.), *Creating Food Futures: Trade, Ethics, and the Environment*. London, Routledge

Richardson, J. (2008) 'Is your food really organic?' *Alternet.org*

Rifkin, J. (1998) *The Biotech Century: How Genetic Commerce Will Change the World*. London, Weidenfeld and Nicholson

Rochon, T. R. (1998) *Culture Moves: Ideas, Activism, and Changing Values*. Princeton, Princeton University Press

Rose, N. (1999) *Powers of Freedom – Reframing Political Thought*. Cambridge, Cambridge University Press

Ross, A. (1994) *The Chicago Gangster Theory of Life: Nature's Debt to Society*. London, Verso

Rowell, A. (1996). *Green Backlash: Global Subversion of Environmental Movement*. London, Routledge

Sanderson, W. (1929) *Statecraft*. London, Methuen and Co

Sanderson, W. (1933) *The English Mistery*. London

Schmitt, M. (2006) 'Fertile minds and friendly pens: Early women pioneers', in G. Holt and M. Reed (eds), *Sociological Perspectives on Organic Agriculture*. Wallingford, CABI, pp56–70

Schumacher, E. F. (1971) 'Soil Association – Aims', *SPAN*, October, p316

Schumacher, E. F. (1974) *Small Is Beautiful. A Study of Economics as if People Mattered*. London, Abacus

Scmit, J. (2008) 'Organic food sales feel the bite from sluggish economy', *USA Today*

Scoones, I. (2008) 'Mobilizing against GM crops in India, South Africa and Brazil', *Journal of Agrarian Change*, vol 8, nos 2–3, pp315–344

Scott Williamson, G. and I. Pearse (1947) *Biologists in Search of Material: An Interim Report on the Work of the Pioneer Health Centre Peckham*. London, Faber and Faber

Seel, B., M. Paterson and B. Doherty (eds.) (2000) *Direct Action in British Environmentalism*. London, Routledge

Shiva, V. (1991) *The Violence of the Green Revolution: Ecological Degradation and Political Conflict*. Zed Books, London

Shulman, S. W. (1999a) 'The business of soil fertility: A convergence of urban–agrarian concern in the early 20th century', *Organization and Environment*, vol 12, no 4, pp401–424

Shulman, S. W. (1999b) 'The progressive era farm press: A primer on a neglected source of journalism history', *Journalism History*, vol 25, no 1, pp27–35

Sligh, M. and T. Cierpaka (2007) 'Organic values', in W. Lockeretz (ed.), *Organic Farming. An International History*. Wallingford, CABI

Smith, E. and T. Marsden (2004) 'Exploring the "limits to growth" in UK organics: Beyond the statistical image', *Journal of Rural Studies*, vol 20, pp345–357

Soil Association (1946) 'An introduction to the Soil Association', *Mother Earth*, vol 1

Soil Association (1998) *The Organic Food and Farming Report 1998*. Bristol, The Soil Association

Soil Association (1999) *A Report on the Dispersal of Maize Pollen*. Bristol, The Soil Association

Soil Association (2006) *The Organic Market Report 2006*. Bristol, The Soil Association

Soil Association (2009) *The Organic Market Report 2009*, Bristol, The Soil Association

Staff-IBM (1952) 'DDT', *Information Bulletin for Members*, no 52, p6

Staff-JSA (1967) 'Our new friends', *Journal of the Soil Association*, pp433–435

Staff-ME (1963) 'Obituary Jorian Jenks', *Mother Earth*

Staff (1968) 'The Soil Association buys farms', *The Times*,

Stone, D. (1999) 'The extremes of Englishness: The "exceptional" ideology of Anthony Mario Ludovici', *Journal of Political Ideologies*, vol 4, no 2, pp191–218

Stone, D. (2003) 'The English mistery, the BUF, and the dilemmas of British fascism', *The Journal of Modern History*, vol 75, pp336–358

Swann, M. M. (1969) *Report of the Joint Committee on the Use of Antibiotics in Animal Husbandry and Veterinary Medicine*. London, HMSO

Sykes, F. (1946) 'Our Association', *Mother Earth*, p9

Sykes, F. (1959) *Modern Humus Farming*. London, Faber and Faber

Tansey, G. and J. D'Silva (eds) (1999) *The Meat Business: Devouring a Hungry Planet*. London, Earthscan

Tarrow, S. (1996) *Power in Movement*. Cambridge, Cambridge University Press

Thrift, N. (2004) *Knowing Capitalism*. London, Sage

Throckmorton, R. I. (1952) 'Organic farming – Bunk!' *Reader's Digest*, October

Thurlow, R. (2006) *Fascism in Britain. From Oswald Mosley's Blackshirts to the National Front*. London, I. B. Tauris

Tilly, C. (2004) *Social Movements 1768–2004*. London, Paradigm Publishers

Toke, D. (2002a) 'GM crops: Science, Policy and Environmentalists'. Birmingham, Department of Political Science and International Affairs, University of Birmingham, p10

Toke, D. (2002b) 'UK GM crop policy – Relative calm before the storm?' Birmingham, Department of Political Science and International Affairs, p21

Tovey, H. (1997) 'Food, environmentalism and rural sociology: On the organic farming movement in Ireland', *Sociologia Ruralis*, vol 37, no 1

Trentman, F. (2007) 'Citizenship and consumption', *Journal of Consumer Culture*, vol 7, no 2, pp147–158

Tsoukas, H. (1999) 'David and Goliath in the risk society: Making sense of the conflict between Shell and Greenpeace in the North Sea', *Organization*, vol 6, no 3, pp499–528

Turner, D. (2000) Jorian Jenks. M. Reed personal communication.

Twinch, C. (2001) *Tithe War 1918–1939*. London, Media Associates

Vidal, J. (1997) *McLibel: Burger Culture on Trial*. London, Pan Books

Vidal, J. (1999) 'How Monsanto's mind was changed' *The Guardian*, 9 October 1999

Voelcker, A. (1883) *A Report on the Condition of Indian Agriculture (sic)*. London, Eyre and Spottiswoode

Vogt, G. (2007) 'The origins of organic farming', in W. Lockeretz (ed.), *Organic Farming: An International History*. Wallingford, CABI, pp9–30

Waddell, C. (ed.) (2000) *And No Birds Sing: Rhetorical Analysis of Rachel Carson's Silent Spring*. Illinois, Southern Illinois University Press

Wall, D. (1999) *Earth First*. London, Routledge

Waller, B. (1962) *Prophet of the New Age*. London, Faber and Faber

Waller, B. (1963) Thanks. Rolf Gardiner Archive

Waller, B. (1966 (approx.)) Reg Milton. Rolf Gardiner Archive

Wallop, H. (2007) 'Organic air-freight food to be stripped of status', *The Daily Telegraph*

Webb, J. (1976) *The Occult Underground*. London, Open Court Publishing

Webber, G. C. (1986) *The Ideology of the British Right 1918–1939*. London, Croom Helm

Whole Earth Foods (2009) www.wholeearthfoods.com, last accessed 14 March 2009

Wilson, C. D. and Lord Kitchener (1960) 'Wholefood', *Mother Earth*, vol 14, p104

Wilson, C. D. and Lord Kitchener (1963) 'The Organic Food Society', *Mother Earth*, p105

Winston, M. L. (1997) *Nature Wars: People vs Pests*. London, Harvard University Press

Winter, D. M. and C. Morris (2002) 'Organic farming and the environment', in I. Douglas (ed.), *Causes and Consequences of Global Environmental Change*. Chichester, John Wiley & Sons, vol 3, pp532–535

Winter, D. M. (2003) 'Embeddness, the new food economy and defensive localism', *Journal of Rural Studies*, vol 19, pp23–32

Wood, D. B. (2006) '*E. coli* cases prompt calls to regulate farm practices', *Christian Science Monitor*

Woods, A. (2004) *A Manufactured Plague. The History of Food-and-Mouth Disease in Britain*. London, Earthscan

Woodward, L. (2004) 'Mary Langman pioneering spirit who helped launch the wholefood movement' *The Guardian*

Wrench, G. T. (1938) *The Wheel of Health*. London, Daniels

Wrench, G. T. (1939) *The Restoration of the Peasantries*. London, Daniels

Wrench, G. T. (1946) *Reconstruction by Way of the Soil*. London, Faber and Faber

Wrench, G.T. (1972) *The Wheel of Health*. New York, Schocken Books

Wright, J. (2009) *Sustainable Agriculture and Food Security in an Era of Oil Scarcity: Lessons from Cuba*. London, Earthscan

Wright, P. (1996) *The Village that Died for England – The Strange Case of Tyneham*. London, Cape

Index